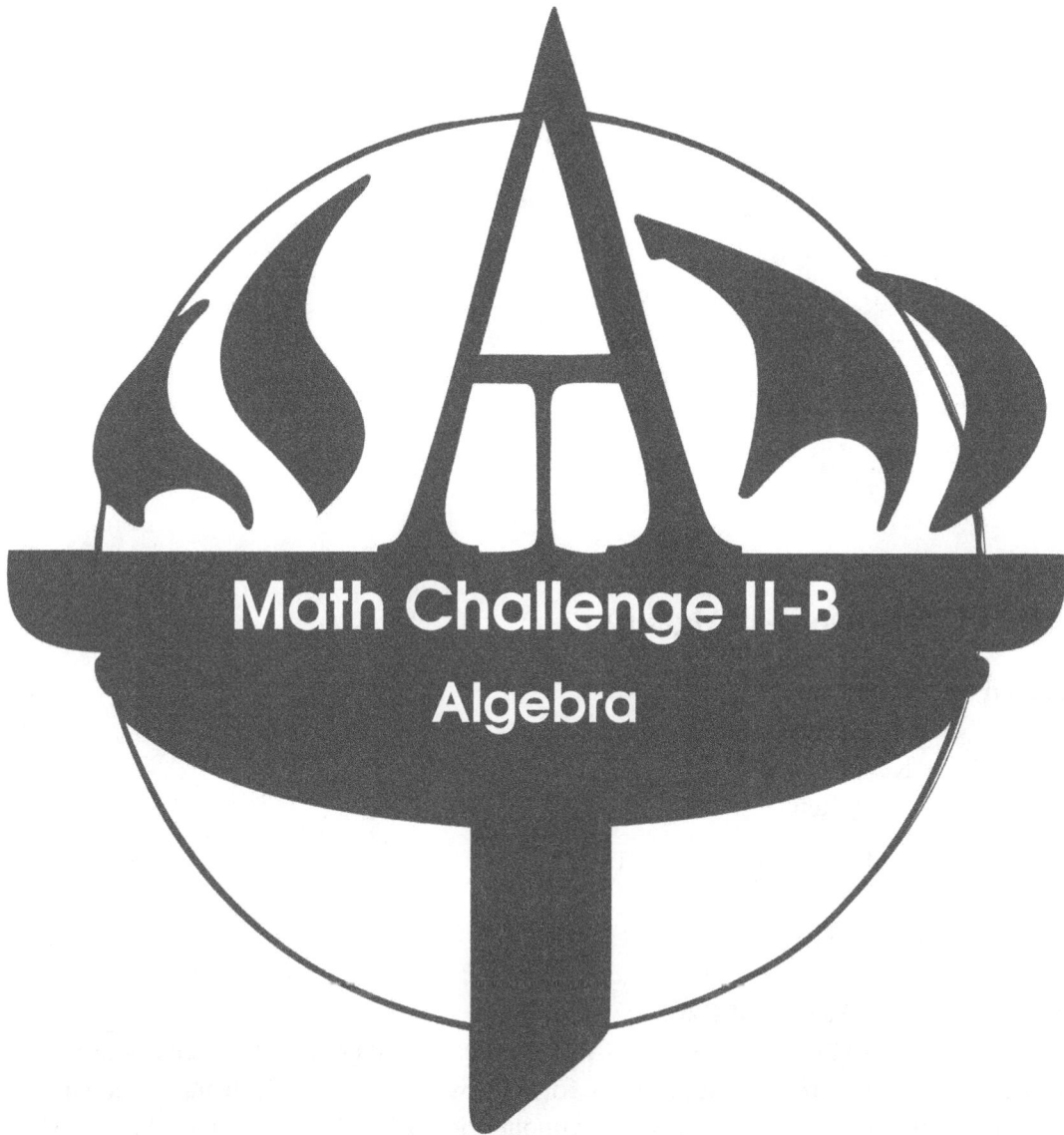

Math Challenge II-B
Algebra

Areteem Institute

Math Challenge II-B Algebra

Edited by   John Lensmire
David Reynoso
Kevin Wang
Kelly Ren

ISBN: 1-944863-25-7
ISBN-13: 978-1-944863-25-8

First printing, March 2019.

Math Challenge II-A Combinatorics
Math Challenge II-B Combinatorics
Math Challenge III Combinatorics
Math Challenge I-A Number Theory
Math Challenge I-B Number Theory
Math Challenge I-C Finite Math
Math Challenge II-A Number Theory
Math Challenge II-B Number Theory
Math Challenge III Number Theory

COMING SOON FROM ARETEEM PRESS

Fun Math Problem Solving For Elementary School Vol. 2 (and Solutions Manual)
Counting & Probability for Middle School (and Solutions Manual) - From Common Core to Math Competitions
Number Theory Problem Solving for Middle School (and Solutions Manual) - From Common Core to Math Competitions

The books are available in paperback and eBook formats (including Kindle and other formats).
To order the books, visit https://areteem.org/bookstore.

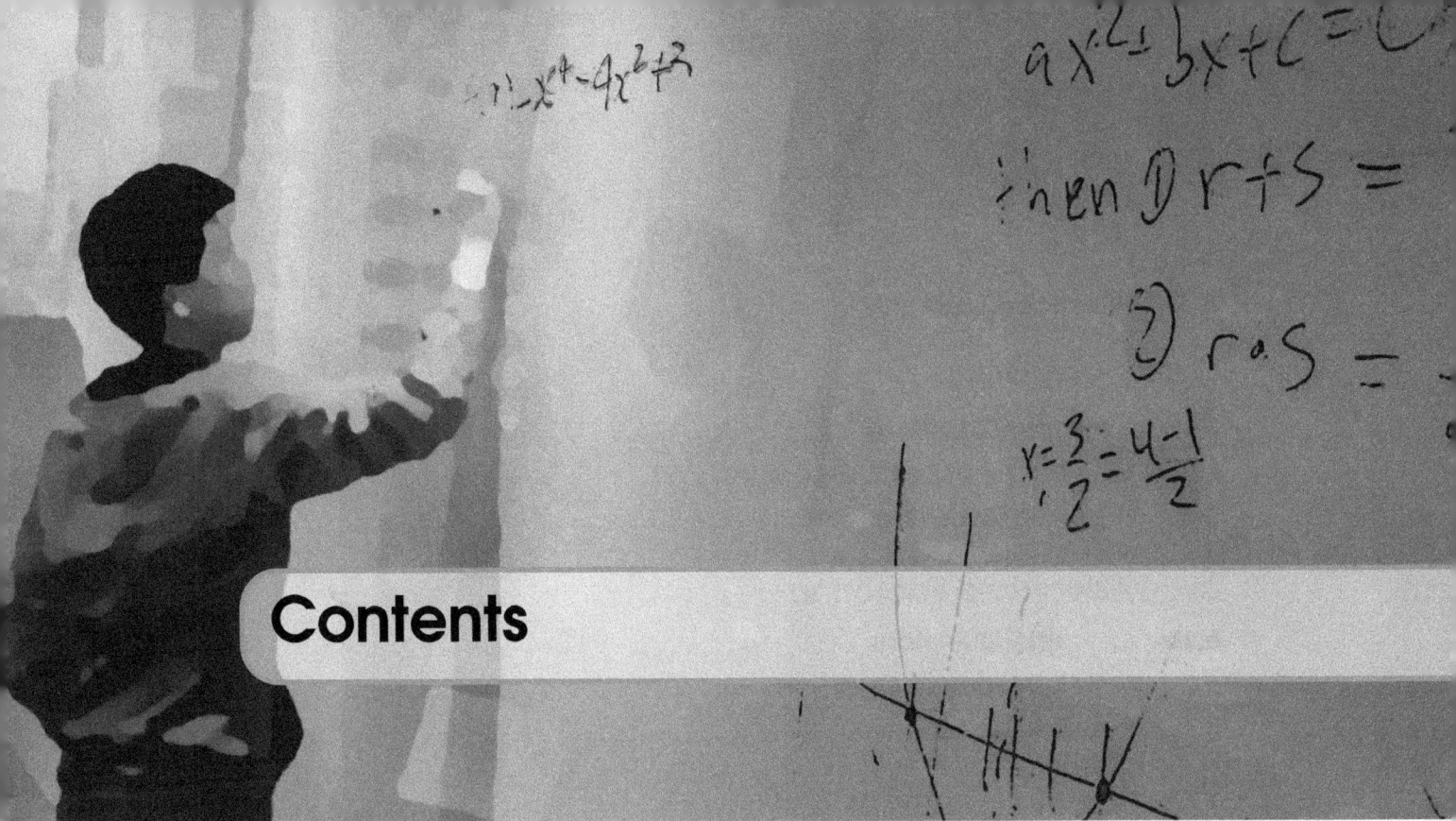

Contents

Introduction 9

1 Factoring 15

1.1 Example Questions 16

1.2 Quick Response Questions 19

1.3 Practice Questions 22

2 Quadratics 25

2.1 Example Questions 26

2.2 Quick Response Questions 30

2.3 Practice Questions 32

3 Complex Numbers I 35

3.1 Example Questions 36

3.2 Quick Response Questions 39

3.3 Practice Questions 41

4 Complex Numbers II 43

4.1 Example Questions 44

4.2 Quick Response Questions 47

| 4.3 | Practice Questions | 50 |

5	**Polynomials**	**53**
5.1	Example Questions	55
5.2	Quick Response Questions	58
5.3	Practice Questions	60

6	**Advanced Factoring**	**63**
6.1	Example Questions	63
6.2	Quick Response Questions	66
6.3	Practice Questions	69

7	**Solving Equations I**	**71**
7.1	Example Questions	72
7.2	Quick Response Questions	75
7.3	Practice Questions	77

8	**Solving Equations II**	**79**
8.1	Example Questions	79
8.2	Quick Response Questions	82
8.3	Practice Questions	84

9	**Exponents and Logarithms**	**87**
9.1	Example Questions	88
9.2	Quick Response Questions	91
9.3	Practice Questions	93

Solutions to the Example Questions		**95**
1	Solutions to Chapter 1 Examples	96
2	Solutions to Chapter 2 Examples	102
3	Solutions to Chapter 3 Examples	108
4	Solutions to Chapter 4 Examples	115
5	Solutions to Chapter 5 Examples	122
6	Solutions to Chapter 6 Examples	127

7   **Solutions to Chapter 7 Examples** 132

8   **Solutions to Chapter 8 Examples** 137

9   **Solutions to Chapter 9 Examples** 142

Introduction

In Math Challenge II-B, students learn and practice in areas such as algebra and geometry at the high school level, as well as advanced number theory and combinatorics. Topics include polynomials, inequalities, special algebraic techniques, trigonometry, triangles and polygons, collinearity and concurrency, vectors and coordinates, numbers and divisibility, modular arithmetic, residue classes, advanced counting strategies, binomial coefficients, pigeonhole principle, sequence and series, and various other topics and problem solving techniques involved in math contests such as the American Mathematics Competition (AMC) 10/12, ARML, and also beginning AIME.

The course is divided into four terms:

- Summer, covering Algebra
- Fall, covering Geometry
- Winter, covering Combinatorics
- Spring, covering Number Theory

The book contains course materials for Math Challenge II-B: Algebra.

We recommend that students take all four terms, but the terms do not build on previous terms, so they do not need to be taken in order and students can take single terms if they want to focus on specific topics.

Students can sign up for the online live or self-paced course at classes.areteem.org.

About Areteem Institute

Areteem Institute is an educational institution that develops and provides in-depth and advanced math and science programs for K-12 (Elementary School, Middle School, and High School) students and teachers. Areteem programs are accredited supplementary programs by the Western Association of Schools and Colleges (WASC). Students may attend the Areteem Institute in one or more of the following options:

- Live and real-time face-to-face online classes with audio, video, interactive online whiteboard, and text chatting capabilities;
- Self-paced classes by watching the recordings of the live classes;
- Short video courses for trending math, science, technology, engineering, English, and social studies topics;
- Summer Intensive Camps held on prestigious university campuses and Winter Boot Camps;
- Practice with selected free daily problems and monthly ZIML competitions at ziml.areteem.org.

Areteem courses are designed and developed by educational experts and industry professionals to bring real world applications into STEM education. The programs are ideal for students who wish to build their mathematical strength in order to excel academically and eventually win in Math Competitions (AMC, AIME, USAMO, IMO, ARML, MathCounts, Math Olympiad, ZIML, and other math leagues and tournaments, etc.), Science Fairs (County Science Fairs, State Science Fairs, national programs like Intel Science and Engineering Fair, etc.) and Science Olympiads, or for students who purely want to enrich their academic lives by taking more challenging courses and developing outstanding analytical, logical, and creative problem solving skills.

Since 2004 Areteem Institute has been teaching with methodology that is highly promoted by the new Common Core State Standards: stressing the conceptual level understanding of the math concepts, problem solving techniques, and solving problems with real world applications. With the guidance from experienced and passionate professors, students are motivated to explore concepts deeper by identifying an interesting problem, researching it, analyzing it, and using a critical thinking approach to come up with multiple solutions.

Thousands of math students who have been trained at Areteem have achieved top honors and earned top awards in major national and international math competitions, including Gold Medalists in the International Math Olympiad (IMO), top winners and qualifiers at the USA Math Olympiad (USAMO/JMO) and AIME, top winners at the

Zoom International Math League (ZIML), and top winners at the MathCounts National Competition. Many Areteem Alumni have graduated from high school and gone on to enter their dream colleges such as MIT, Cal Tech, Harvard, Stanford, Yale, Princeton, U Penn, Harvey Mudd College, UC Berkeley, or UCLA. Those who have graduated from colleges are now playing important roles in their fields of endeavor.

Further information about Areteem Institute, as well as updates and errata of this book, can be found online at `http://www.areteem.org`.

About Zoom International Math League

The Zoom International Math League (ZIML) has a simple goal: provide a platform for students to build and share their passion for math and other STEM fields with students from around the globe. Started in 2008 as the Southern California Mathematical Olympiad, ZIML has a rich history of past participants who have advanced to top tier colleges and prestigious math competitions, including American Math Competitions, MATHCOUNTS, and the International Math Olympaid.

The ZIML Core Online Programs, most available with a free account at `ziml.areteem.org`, include:

- **Daily Magic Spells:** Provides a problem a day (Monday through Friday) for students to practice, with full solutions available the next day.
- **Weekly Brain Potions:** Provides one problem per week posted in the online discussion forum at `ziml.areteem.org`. Usually the problem does not have a simple answer, and students can join the discussion to share their thoughts regarding the scenarios described in the problem, explore the math concepts behind the problem, give solutions, and also ask further questions.
- **Monthly Contests:** The ZIML Monthly Contests are held the first weekend of each month during the school year (October through June). Students can compete in one of 5 divisions to test their knowledge and determine their strengths and weaknesses, with winners announced after the competition.
- **Math Competition Practice:** The Practice page contains sample ZIML contests and an archive of AMC-series tests for online practice. The practices simulate the real contest environment with time-limits of the contests automatically controlled by the server.
- **Online Discussion Forum:** The Online Discussion Forum is open for any comments and questions. Other discussions, such as hard Daily Magic Spells or the Weekly Brain Potions are also posted here.

These programs encourage students to participate consistently, so they can track their progress and improvement each year.

In addition to the online programs, ZIML also hosts onsite Local Tournaments and Workshops in various locations in the United States. Each summer, there are onsite ZIML Competitions at held at Areteem Summer Programs, including the National ZIML Convention, which is a two day convention with one day of workshops and one day of competition.

ZIML Monthly Contests are organized into five divisions ranging from upper elementary school to advanced material based on high school math.

- **Varsity:** This is the top division. It covers material on the level of the last 10 questions on the AMC 12 and AIME level. This division is open to all age levels.
- **Junior Varsity:** This is the second highest competition division. It covers material at the AMC 10/12 level and State and National MathCounts level. This division is open to all age levels.
- **Division H:** This division focuses on material from a standard high school curriculum. It covers topics up to and including pre-calculus. This division will serve as excellent practice for students preparing for the math portions of the SAT or ACT. This division is open to all age levels.
- **Division M:** This division focuses on problem solving using math concepts from a standard middle school math curriculum. It covers material at the level of AMC 8 and School or Chapter MathCounts. This division is open to all students who have not started grade 9.
- **Division E:** This division focuses on advanced problem solving with mathematical concepts from upper elementary school. It covers material at a level comparable to MOEMS Division E. This division is open to all students who have not started grade 6.

To participate in the ZIML Online Programs, create a free account at ziml.areteem. org. The ZIML site features are also provided on the ZIML Mobile App, which is available for download from Apple's App Store and Google Play Store.

Acknowledgments

This book contains many years of collaborative work by the staff of Areteem Institute. This book could not have existed without their efforts. Huge thanks go to the Areteem staff for their contributions!

The examples and problems in this book were either created by the Areteem staff or adapted from various sources, including other books and online resources. Especially, some good problems from previous math competitions and contests such as AMC, AIME, ARML, MATHCOUNTS, and ZIML are chosen as examples to illustrate concepts or problem-solving techniques. The original resources are credited whenever possible. However, it is not practical to list all such resources. We extend our gratitude to the original authors of all these resources.

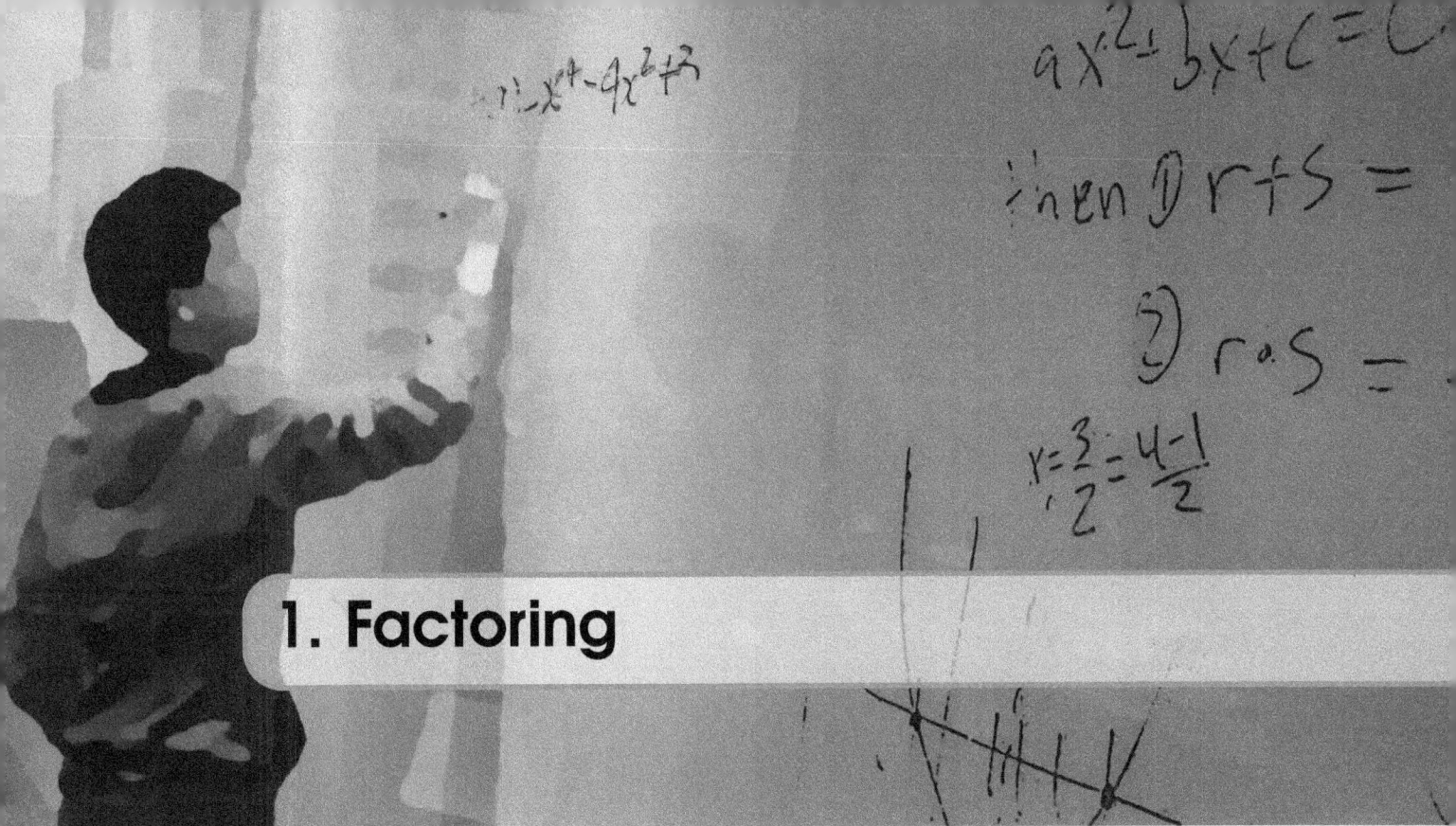

1. Factoring

Factoring Basics

- **Finding Common Factors**: The first step in any factoring problem is to search for common factors. Factoring out the common factor is the backward application of the Distributive Law: $ab + ac = a(b+c)$. Here a, b, c can be numbers or expressions. Sometimes it takes some work in order to get the common factor to appear.
- **Commonly used formulas**:

$$a^2 - b^2 = (a+b)(a-b)$$
$$a^2 \pm 2ab + b^2 = (a \pm b)^2$$
$$a^3 \pm 3a^2b + 3ab^2 \pm b^3 = (a \pm b)^3$$
$$a^3 + b^3 = (a+b)(a^2 - ab + b^2)$$
$$a^3 - b^3 = (a-b)(a^2 + ab + b^2)$$
$$a^2 + b^2 + c^2 + 2ab + 2bc + 2ca = (a+b+c)^2$$
$$a^3 + b^3 + c^3 - 3abc = (a+b+c)(a^2+b^2+c^2 - ab - bc - ca)$$
$$a^n - b^n = (a-b)(a^{n-1} + a^{n-2}b + a^{n-3}b^2 + \cdots + ab^{n-2} + b^{n-1}), \quad n \in \mathbb{N}$$
$$a^n - b^n = (a+b)(a^{n-1} - a^{n-2}b + a^{n-3}b^2 - \cdots + ab^{n-2} - b^{n-1}), \quad n \in \mathbb{N} \text{ even}$$
$$a^n + b^n = (a+b)(a^{n-1} - a^{n-2}b + a^{n-3}b^2 - \cdots - ab^{n-2} + b^{n-1}), \quad n \in \mathbb{N} \text{ odd}$$

- **Cross-multiplication**: Cross-multiplication is the reverse of the FOIL method: $abx^2 + (ad+bc)x + cd = (ax+c)(bx+d)$.

Further Factoring Techniques

- **Grouping**: If there are 4 or more terms, we can group some terms together and

---

factor the portions first, and sometimes the portions end up with common factors or fit into a formula, then factoring is possible.

- **Split and add terms**: Split one term to two or more terms; add two terms that cancel each other. These techniques can enable the polynomial to be factored by grouping.

1.1   Example Questions

Problem 1.1 Factor the following.

(a) $10x^{10}y^8 + 5x^5y^9$.

(b) $(2x + 3y)^3 - 8x^3 - 27y^3$.

Problem 1.2 Prove the following identities:

(a) Prove the formula for factoring $a^3 + b^3 + c^3 - 3abc$.

(b) Prove the formula for factoring $y^6 - y^3$.

Problem 1.3 Factor $x^{15} + x^{14} + x^{13} + \cdots + x^2 + x + 1$.

Problem 1.4 Factor the following.

(a) Factor $22y^2 - 35y + 3$.

(b) Factor $2x^2 - 7x + 3$.

(c) Factor $3k^2 - 5k - 2$.

Problem 1.5 Factor $a^4 + a^3 + a^2b + ab^2 + b^3 - b^4$.

Problem 1.6 Factor $x^3 - 9x + 8$ by using the given technique:

(a) $8 = -1 + 9$.

(b) $-9x = -x - 8x$.

(c) $x^3 = 9x^3 - 8x^3$.

(d) Add two terms: $-x^2 + x^2$.

Problem 1.7 Factor the following.

(a) $x^9 + x^6 + x^3 - 3$.

(b) (Sophie Germain's technique) $x^4 + 4y^4$.

Problem 1.8 Factor the expression $(x+1)^4 + (x^2-1)^2 + (x-1)^4$.

Problem 1.9 Factor $a^2 + b^2 + c^2 - 2bc + 2ca - 2ab$.

Problem 1.10 Factor $-2x^{5n-1}y^n + 4x^{3n-1}y^{n+2} - 2x^{n-1}y^{n+4}$.

1.2 Quick Response Questions

Problem 1.11 What is the greatest common factor of the terms in $26x^3yz^3 + 91x^2y^7z^5$?

(A) $91x^3y^7z^5$
(B) $13x^2yz^3$
(C) $117x^5y^8z^8$
(D) $2x + 7y^6z^2$

Problem 1.12 Factor $1 + x + y + xy$.

(A) $x(1 + y)$
(B) $(x + y)(x + y)$
(C) $(1 + x)(1 + y)$
(D) $(y + x)(x + 1)$

Problem 1.13 $x^2 + 2xy + y^2 - 9$ factors as $(x + y - k)(x + y + k)$. What is k?

Problem 1.14 Factor $x^3 + x^2 - 4x - 4$.

(A) $x^2(x - 4)$
(B) $(x - 2)(x + 1)(x - 1)$
(C) $x^3(1 - 4x)$
(D) $(x + 2)(x - 2)(x + 1)$

Problem 1.15 Factor $1 + x + y + z + xy + xz + yz + xyz$.

(A) $(1+x)(1+y)(1+z)$
(B) $z(1+x+y+xy)$
(C) $(x+y+z)(1+xyz)$
(D) $(1+x)(y+z)$

Problem 1.16 Factor $x^2 - 6x + 8$ using $8 = 9 - 1$.

(A) $(x+9)(x-1)$
(B) $(x+3)(x-3-1)$
(C) $(x-4)(x-2)$
(D) $(x+4)(x-2)$

Problem 1.17 Factor $2x^2 + 7x + 6$.

(A) $(2x+1)(x+3)$
(B) $(2x+3)(x+2)$
(C) $(x+3)(2x+2)$
(D) $(2x+1)(x+6)$

Problem 1.18 Factor $6x^2 + x - 35$.

(A) $(2x-5)(3x+7)$
(B) $(6x+1)(x-35)$
(C) $(2x+5)(3x-7)$
(D) $(3x+5)(2x-7)$

Problem 1.19 Factor $x^4 - 5x^2 + 4$.

(A) $(x^2 + 5)(x^2 - 1)$
(B) $(x - 1)(x + 1)(x - 2)(x + 2)$
(C) $(x - 1)(x - 1)(x + 2)(x + 2)$
(D) $(x^2 + 5)(x + 1)(x - 1)$

Problem 1.20 Factor $81x^4 - 16$.

(A) $(9x^2 + 4)(9x^2 - 4)$
(B) $(3x - 2)(3x - 2)(3x + 2)(3x + 2)$
(C) $(x - 2)(x + 2)(9x - 2)(9x + 2)$
(D) $(3x - 2)(3x + 2)(9x^2 + 4)$

1.3 Practice Questions

Problem 1.21 Factor the following:

(a) Factor $10x^2y^2 - 15xy^3 + 25xy^2z$.

(b) Factor $6x(a-b)^4 - 30x(b-a)^3$.

Problem 1.22 Factor $x^3 - 8y^3 - z^3 - 6xyz$.

Problem 1.23 Factor $a^{32} - b^{32}$.

Problem 1.24 Factor the following:

(a) Factor $2p^2 + p - 3$.

(b) Find all positive n such that $x^2 - nx - 12$ factors.

Problem 1.25 Factor $xy + xz + yw + wz$.

Problem 1.26 Factor $x^3 + x^2 + 4$

Problem 1.27 $a^3b - ab^3 + a^2 + b^2 + 1$.

Problem 1.28 $(m^2 - 1)(n^2 - 1) + 4mn$.

Problem 1.29 Factor $1 + 2x^2 + 2y^2 + 2x^2y^2 + x^4 + y^4$.

Problem 1.30 Factor $a^7 - a^5b^2 + a^2b^5 - b^7$.

2. Quadratics

Quadratic Equations

- A quadratic equation is an equation in x of the form $ax^2 + bx + c = 0$ where $a \neq 0$.
- If the coefficients a, b, c are given, the roots (solutions) are determined as well.
- On the other hand, if the roots are given, the coefficients a, b, c are also determined up to a common factor (that is to say, if a, b, c are multiplied by the same nonzero number, the roots don't change).
- This shows that in a quadratic equation, the roots and the coefficients have a very close relationship. This relationship is expressed in the discriminant and Vieta's Theorem.

Roots and the Discriminant

- **Roots of A Quadratic Equation**
    - There are several ways to solve a quadratic equation, such as factoring, completing the square, or the quadratic formula.
    - In general, the roots x_1 and x_2 are given by the quadratic formula:

    $$x_{1,2} = \frac{-b \pm \sqrt{b^2 - 4ac}}{2a}.$$

    - You should know how to derive the quadratic formula.
- **Vertex of a Quadratic**
    - The maximum or minimum value of a quadratic function occurs at the *vertex*.

○ The x-coordinate of the vertex is $\dfrac{-b}{2a}$. Note this is the average of the roots of the quadratic.

- **The Discriminant**
  ○ Note that the expression $b^2 - 4ac$ is inside the square root. Therefore the solution for the quadratic equation depends on the sign of this expression.
  ○ Denote $\Delta = b^2 - 4ac$. This is called the *discriminant*. There are three possibilities:
    1. If $\Delta > 0$, the equation has two distinct real roots.
    2. If $\Delta = 0$, the equation has exactly one real root, or we say it has two identical roots, or double roots.
    3. If $\Delta < 0$, the equation has no real roots.

Vieta's Theorem

- Vieta's Theorem (or Vieta's Formulas) shows the relation between the roots and the coefficients.
- If x_1 and x_2 are the roots for the quadratic equation $ax^2 + bx + c = 0 (a \neq 0)$, then

$$x_1 + x_2 = -\frac{b}{a}, \quad \text{and} \quad x_1 x_2 = \frac{c}{a}.$$

- These relations can be easily proved using the quadratic formula.
- In the AMC 10 and 12, many problems can be easily solved using discriminant or Vieta's Theorem.

2.1   Example Questions

Problem 2.1 Answer the following using the discriminant.

(a) For what values of m does the equation $4x^2 + 8x + m = 0$ have two distinct real roots?

(b) Given that the equation $x^2 - 2x - m = 0$ has no real roots, how many real roots does the equation $x^2 + 2mx + 1 + 2(m^2 - 1)(x^2 + 1) = 0$ have?

Problem 2.2 Use Vieta's formula to solve the following.

(a) A quadratic equation has two roots $\frac{2}{3}$ and $-\frac{1}{2}$, what is this equation? (multiple answers are possible)

(b) Two real numbers have sum -10 and product -5, find these two numbers.

(c) (2005 AMC 10B/12B) The quadratic equation $x^2 + mx + n = 0$ has roots that are twice those of $x^2 + px + m = 0$, and none of m, n, and p is zero. What is the value of n/p?

Problem 2.3 Let x_1, x_2 be the two roots for equation $x^2 + x - 3 = 0$, find the value of $x_1^3 - 4x_2^2 + 19$.

Problem 2.4 Given the equation in x, $x^2 + 2mx + m + 2 = 0$:

(a) For what values of m does the equation have two (not necessarily distinct) positive roots?

(b) For what values of m does the equation have one positive root and one negative root?

Problem 2.5 (2005 AMC 10A/12A) There are two values of a for which the equation $4x^2 + ax + 8x + 9 = 0$ has only one solution for x. What is the sum of those values of a?

Problem 2.6 Find k in each of the following scenarios:

(a) In the equation $x^2 - 402x + k = 0$, one of the roots plus three equals 80 times the other root.

(b) Let x_1 and x_2 be the two roots of the equation $4x^2 - 8x + k = 0$. Suppose further that $\dfrac{1}{x_1} + \dfrac{1}{x_2} = \dfrac{8}{3}$.

Problem 2.7 Do the following.

(a) The sum of squares of the roots of equation $x^2 + 2kx = 3$ is 10. Find the possible values of k.

(b) For equation $2x^2 + mx - 2m + 1 = 0$, the sum of squares of the two real roots is $\dfrac{29}{4}$. Find the value of m.

Problem 2.8 The two real roots of $x^2 + (m-2)x + 5 - m = 0$ are both greater than 2. Find the possible range of values for real number m.

Problem 2.9 If x_1 and x_2 are integer roots of equation $x^2 + mx + 2 - n = 0$, and $(x_1^2 + 1)(x_2^2 + 1) = 10$, how many possible pairs (m, n) are there?

Problem 2.10 Let x_1, x_2 be two positive integer roots of equation $x^2 + px + 1997 = 0$. Find the value of $\dfrac{p}{(x_1 + 1)(x_2 + 1)}$.

2.2 Quick Response Questions

Problem 2.11 Factor $x^2 + 6x + 8$. What is the smallest root of $x^2 + 6x + 8 = 0$.

Problem 2.12 Factor $x^2 + (m+2)x + 2m$. There is an integer L such that L is a root of $x^2 + (m+2)x + 2m = 0$ for all m. What is L?

Problem 2.13 What is the discriminant of the quadratic $3x^2 + 3x + 4$?

Problem 2.14 The roots of $x^2 - 3x + 1 = 0$ are of the form $\dfrac{A \pm \sqrt{B}}{2}$. What is B?

Problem 2.15 For $m = \pm\sqrt{K}$, $2x^2 + mx + 3 = 0$ has exactly one real root. What is K?

Problem 2.16 What is the x-coordinate of the vertex of the quadratic $4x^2 + 24x + 48$?

Problem 2.17 The vertex of the quadratic $x(x - 2a)$ occurs when $x = 4$. The vertex of the quadratic $(x+a)(x-3a)$ occurs when $x = L$. What is L?

Problem 2.18 For how many integers x is $f(x) = x^2 + x - 6$ negative?

Problem 2.19 What is the sum of the roots of $49x^2 - 196x + 193$?

Problem 2.20 The quadratic $2x^2 - Bx + C = 0$ has roots 2 and $1/2$. What is C?

2.3 Practice Questions

Problem 2.21 Suppose that $ax^2 + bx + c = 0$ has real solutions x_1 and x_2. Find the solutions to the equation $a(x + x_1)^2 + b(x + x_1) + c = 0$.

Problem 2.22 For what values of m does $x^2 - 4x - m = 0$ have no real solutions while $x^2 - 9x + m^2 = 0$ has at least one real solution?

Problem 2.23 Let x_1 and x_2 be the two roots of $17x^2 - 8x - 2 = 0$. Use Vieta's Theorem to find $x_1^2 + x_2^2$.

Problem 2.24 Find all m such that $x^2 - (2m - 3)x + m(m - 3) = 0$ has one positive and one negative root.

Problem 2.25 Find all a such that $(a + 1)x^2 + (a - 1)x - 2 = 0$ has exactly one real solution.

Problem 2.26 If x_1 and x_2 are the two real roots of $x^2 + x + q = 0$, and $|x_1 - x_2| = q$, find the value of q.

Problem 2.27 For the equation $x^2 + mx + n = 0$, the difference between the two roots is p and the product of the two roots is q. What is $m^2 + n^2$ in terms of p and q?

Problem 2.28 The quadratic equation $x^2 + 2kx + 2k^2 - 1 = 0$ has at least one negative root. Find the possible range of values for k.

Problem 2.29 Find all ordered pairs (a, b) such that $a^2 + b^2$ is prime, and the equation $x^2 + ax + 1 = b$ has two positive integer roots.

Problem 2.30 The equation $x^2 + (a - 6)x + a = 0$ has two integer roots. Find the value of a.

3. Complex Numbers I

The Complex Numbers

- The number $i = \sqrt{-1}$ is the standard imaginary unit, so $i^2 = -1$.
- A number is a complex number if it can be written as $a + b \cdot i$, where a, b are real numbers.
  - a, b are respectively referred to as the real part and imaginary part. If $z = a + bi$ we write $\mathrm{Re}(z) = a$ and $\mathrm{Im}(z) = b$.
  - The *conjugate* of $z = a + bi$ is $\bar{z} = a - bi = \mathrm{Re}(z) - i \cdot \mathrm{Im}(z)$.
  - The *modulus* of $z = a + bi$, is $|z| = \sqrt{a^2 + b^2}$.
- It is often useful to view the complex numbers in the *complex plane*, the 2D plane with x-axis the real part and the y-axis the imaginary part of the complex number.

Forms of a Complex Number

- A complex number z is written in *rectangular* form when it is expressed as $z = a + bi$.
- It is written in *polar* form, when it is expressed as $z = r(\cos(\theta) + i\sin(\theta))$. Here, $r = |z| = \sqrt{a^2 + b^2}$ (the modulus) and $\tan(\theta) = \dfrac{b}{a}$. θ is called the *argument*.

3.1 Example Questions

Problem 3.1 Calculation Practice

(a) $6 + 8i + \overline{4 + 3i}$.

(b) $(3 + 4i)(-2 + 3i)$

(c) $\dfrac{1+i}{2-i}$.

(d) $(1 - i)^4$.

Problem 3.2 Solve the following quadratics over the complex numbers.

(a) $x^2 - 4x + 8 = 0$.

(b) $4x^2 - 12x + 29 = 0$.

Problem 3.3 Prove the following, where z, w are complex numbers.

(a) $\overline{z \cdot w} = \bar{z} \cdot \bar{w}$.

(b) $z\bar{z} = |z|^2$.

(c) $\dfrac{z}{w} = \dfrac{z\bar{w}}{|w|^2}$.

Problem 3.4 Convert between the rectangular and polar forms: write each of these numbers in the other forms.

(a) i

(b) $\sqrt{3} + i$.

(c) $2\cos(180°) + 2i\sin(180°)$.

(d) $\cos(7\pi/6) + i\sin(7\pi/6)$.

Problem 3.5 What are the sets of points satisfying the following? Draw the diagrams.

(a) $|z| \leq 2$.

(b) $\text{Re}(z) > \dfrac{1}{2}$.

(c) $\operatorname{Re}(z) = \operatorname{Im}(z)$.

(d) $\left| \dfrac{z-1}{z+1} \right| < 1$

Problem 3.6 What is $\left(\dfrac{\sqrt{2}}{2}(-1+i) \right)^{100}$?

Problem 3.7 Given that $2 + ai$ and $b + i$ are the two roots of the quadratic equation $x^2 + px + q = 0$ where p and q are real numbers. What are p and q?

Problem 3.8 (2009 AMC 12) For what value of n is $i + 2i^2 + 3i^3 + \cdots + n \cdot i^n = 48 + 49i$?

Problem 3.9 Prove the triangle inequality for complex numbers. That is, for complex numbers x and y, show that
$$|x+y| \le |x| + |y|.$$

Problem 3.10 Suppose z is not real and $|z| = 1$, show that $w = \dfrac{z-1}{z+1}$ is a pure imaginary number.

3.2   Quick Response Questions

Problem 3.11 What is the real part of $(3+4i)(5-2i)$?

Problem 3.12 Find the real part of $(5+i)\overline{(5-i)}$. Round your answer to the nearest tenth if necessary.

Problem 3.13 Find the imaginary part of $\dfrac{2+i}{3-i}$. Express your answer as a decimal.

Problem 3.14 Consider $z = 5+6i$. What is $|z|^2$?

Problem 3.15 If the complex number $3-3i$ is written in polar form, what is the argument θ, with $0° \leq \theta < 360°$?

Problem 3.16 Which of the following is true about $z = 2+4i$?

(A)   The real part of z is 4
(B) $\bar{z} = -2+4i$
(C) $|z|^2 = 2\sqrt{5}$
(D) $|z| = 2\sqrt{5}\bar{z} = -2+4i$

Problem 3.17 Which of the following is not equal to the other 3 expressions?

(A) $2i$

(B) $(1+i)^2$

(C) $\cos(3\pi/2) + i\sin(3\pi/2)$

(D) $2\cos(\pi/2) + 2i\sin(\pi/2)$

Problem 3.18 $2+i$ is a root of $x^2 + Bx + C$. What is B?

Problem 3.19 Let $f(z) = z^4 + z^3 + z^2 + z + 1$. What is the real part of $f(i)$?

Problem 3.20 The roots of $2x^2 + 242 = 0$ are $\pm K \cdot i$ for an integer K. What is K?

3.3 Practice Questions

Problem 3.21 Answer the following:

(a) $\overline{3+4i+\overline{1-5i}}$.

(b) $\dfrac{\overline{4+3i}}{\overline{2-i}}$.

Problem 3.22 Prove that if z is a root of $ax^2 + bx + c$ then so is $\bar{z}$, without using the quadratic formula.

Problem 3.23 Suppose p and q are integers and the quadratic equation $x^2 + px + q = 0$ has two complex roots $2 + ai$ and $b + i$. What is q?

Problem 3.24 Convert the following.

(a) $\sqrt{6} + i\sqrt{2}$ to polar coordinates.

(b) $4(\cos(150°) + i\sin(150°))$.

Problem 3.25 What are the sets of points satisfying the following? Draw the diagrams.

(a) $2 < |z| \le 4$

(b) $\text{Im}(z) \leq \dfrac{1}{2}$

Problem 3.26 Calculate $(1+i)^{102} + (1-i)^{102}$.

Problem 3.27 Given that $a - 3i$ and $c + di$ are the two roots of the quadratic equation $x^2 + px + 12 = 0$. What are the possible values of p?

Problem 3.28 Let $f(z) = 6z^5 + 5z^4 + 4z^3 + 3z^2 + 2z + 1$. Then $f(i) = A + Bi$ for integers A, B. What is $A^2 + B^2$? Note: Here $i = \sqrt{-1}$.

Problem 3.29 Prove that $|z_1 + z_2|^2 + |z_1 - z_2|^2 = 2\left(|z_1|^2 + |z_2|^2\right)$.

Problem 3.30 Suppose that $z = a + bi$ with $|z| = 3$, but $w = \dfrac{z-3}{z+3}$ is a real number. What is $a + b$?

4. Complex Numbers II

Review of Forms of a Complex Number

- A complex number z is written in *rectangular* form when it is expressed as $z = a + bi$.
- It is written in *polar* form, when it is expressed as $z = r(\cos(\theta) + i\sin(\theta))$. Here, $r = |z| = \sqrt{a^2 + b^2}$ (the modulus) and $\tan(\theta) = \dfrac{b}{a}$. θ is called the *argument*.

Multiplication with Polar Coordinates

- If $z = r(\cos(\theta) + i\sin(\theta))$ and $w = s(\cos(\phi) + i\sin(\phi))$, then

$$z \cdot w = (r \cdot s)(\cos(\theta + \phi) + i\sin(\theta + \phi)).$$

  That is, to multiply two complex numbers with polar coordinates, you multiply the moduli and add the arguments.
- In particular, repeating the above formula gives *De Moivre's Formulas*: If $z = r(\cos(\theta) + i\sin(\theta))$ then $z^n = r^n(\cos(n\theta) + i\sin(n\theta))$.
- Note: The proofs of these two results are not the focus of this course.

Roots of Unity

- A complex number is an *nth root of unity* if it is a solution to $z^n = 1$.
- The complex numbers

$$\omega_k = \cos\frac{2k\pi}{n} + i\sin\frac{2k\pi}{n}, \qquad k = 0, 1, 2, \ldots, n-1.$$

are all nth roots of unity.

4.1 Example Questions

Problem 4.1 Review of Rectangular and Polar Form

(a) Convert the following to polar form: $4 - 4i$.

(b) Convert the following to rectangular form: $4(\cos(\pi/3) + i\sin(\pi/3))$.

Problem 4.2 Solve the following quadratics with complex coefficients. Then verify Viete's formulas still hold for the sum and product of the roots.

(a) $2x^2 + ix + 1 = 0$.

(b) $x^2 - 2x + i(3x - 6) = 0$.

Problem 4.3 Perform the following multiplications using both rectangular and polar form to verify the result for multiplying complex numbers in polar form.

(a) $(1 + i) \cdot (1 - i)$.

(b) $\left(\dfrac{1}{2} + i\dfrac{\sqrt{3}}{2} \right) \cdot (2i)$.

Problem 4.4 Do the following:

(a) Write $(1-i)^{40}$ as $a+ib$ with a and b real numbers and simplify your answer.

(b) Write $(\sqrt{3}+i)^8$ in polar and in rectangular form.

Problem 4.5 Roots of Unity Practice

(a) Write the 4th roots of unity in rectangular form.

(b) Write (in rectangular form) the 12th roots of unity with argument (strictly) between $\dfrac{\pi}{2}$ and π.

Problem 4.6 Solve each of the following equations (over the complex numbers) and graph the solutions in the complex plane. If the answer cannot be simplified, it is okay to leave it in polar form.

(a) $x^5 = -32$.

(b) $x^2 = i$.

(c) $x^3 = 8i$.

Problem 4.7 If $|a| = |b| = 1$ and $a + b + 1 = 0$, what are a and b?

Problem 4.8 (2002 AMC 12A) Find the number of ordered pairs of real numbers (a, b) such that $(a + bi)^{2002} = a - bi$. Note: You should also be able to describe all such solutions.

Problem 4.9 Find all the roots (including complex) to $x^4 + x^2 + 1 = 0$. Write your answer in rectangular form.

Problem 4.10 Let $\omega_1 = \cos\dfrac{2\pi}{n} + i\sin\dfrac{2\pi}{n}$ (so ω_1 is an nth root of unity). If a is an nth root of unity, prove that all n solutions to $z^n = 1$ are $a, a \cdot \omega_1, a \cdot (\omega_1)^2, \ldots, a \cdot (\omega_1)^{n-1}$.

4.2 Quick Response Questions

Problem 4.11 What is $-i$ in polar form?

(A) $\cos(\pi/2) + i\sin(\pi/2)$
(B) $\cos(\pi) + i\sin(\pi)$
(C) $\cos(3\pi/2) + i\sin(3\pi/2)$
(D) $\cos(9\pi/4) + i\sin(9\pi/4)$

Problem 4.12 What is the argument of

$$3(\cos(\pi/4) + i\sin(\pi/4)) \times 2(\cos(\pi/5) + i\sin(\pi/5))?$$

Give your answer in degrees, rounded to the nearest degree if necessary.

Problem 4.13 What is $(1 + i\sqrt{3})^7$ in polar form?

(A) $128(\cos(60°) + i\sin(60°))$
(B) $128(\cos(150°) + i\sin(150°))$
(C) $64(\cos(60°) + i\sin(60°))$
(D) $128(\cos(300°) + i\sin(300°))$

Problem 4.14 Factor $x^2 + 3ix + 4$.

(A) $(x+4)(x-1)$
(B) $(x+4i)(x-1)$
(C) $(x+4i)(x-i)$
(D) $(x-4i)(x+i)$

Problem 4.15 Which is NOT an eighth root of unity?

(A) i

(B) $\cos(\pi/3) + i\sin(\pi/3)$

(C) $-\dfrac{\sqrt{2}}{2} - i\dfrac{\sqrt{2}}{2}$

(D) $\cos(3\pi/4) + i\sin(3\pi/4)$

Problem 4.16 What is the sum of all the complex (not real) roots of $x^8 - 1 = 0$? Round your answer to the nearest tenth if necessary.

Problem 4.17 What is the argument of $\sqrt{i}$? Give your answer in degrees.

Problem 4.18 Let z be an $(2n)$th root of unity. Which of the following is true?

(A) z is a nth root of unity

(B) z is a $(4n)$th root of unity

(C) $2z$ is a $(2n)$th root of unity

(D) $\sqrt{z}$ is a $(2n)$th root of unity

Problem 4.19 Consider the equation $ax^2 + bix + c = 0$ where a, b, c are real numbers. If $r + si$ is a root, which of the following must also be a root?

(A) $r - si$

(B) $-r + si$

(C) $s + ri$

(D) None of the above must be roots.

Problem 4.20 Describe all solutions to $z^2 = \bar{z}^2$.

(A) $z = a + bi$ such that $a = 0$.
(B) $z = a + bi$ such that $b = 0$.
(C) $z = a + bi$ such that $a = 0$ and $b = 0$.
(D) $z = a + bi$ such that $a = 0$ or $b = 0$.

4.3   Practice Questions

Problem 4.21 Convert $k + ki$ (for k a real number) to polar coordinates.

Problem 4.22 The quadratic $x^2 + (2i - 1)x + 1 + 5i = 0$ has root $-1 + i$. What is the other root?

Problem 4.23 Consider the sequence $(-2 + 2i)^1$, $(-2 + 2i)^2$, $(-2 + 2i)^3$, etc. What is the first positive real number in this sequence?

Problem 4.24 Find the angle between 0 and 2π that is an argument of $(-2 + 2i\sqrt{3})^{2016}$.

Problem 4.25 Write the 8th roots of unity in rectangular form.

Problem 4.26 Solve the equation $x^4 = 4i$ over the complex numbers.

Problem 4.27 Suppose $|a| = |b| = 2$ and $a + b + 4 = 0$. What are a and b?

Problem 4.28 Suppose n is a positive integer, and z is a complex number with modulus 1, such that z^{2n} is not -1, show that $\dfrac{z^n}{1 + z^{2n}}$ is a real number.

Problem 4.29 (1984 AIME) The equation $z^6 + z^3 + 1$ has complex roots with argument θ between $90°$ and $180°$ in the complex plane. Determine the degree measure of θ.

Problem 4.30 We learned about roots of unity in class. Let $\omega_0 = 1, \omega_1, \omega_2, \ldots, \omega_{11}$ be the 12th roots of unity. For which k in $0, 1, 2, \ldots, 11$ does $\omega_k, (\omega_k)^2, \ldots, (\omega_k)^{12}$ generate all 12 roots of unity?

5. Polynomials

General Definitions

- **Polynomial.** A function that is made of adding multiples of powers of a variable. Examples: $5x^3 - 2x + 1$, $-a^{99} - 9a^9 - 99$, and $4z^2 - 12z + 9$. Usually polynomials are written with the powers going up or going down.
- **Monomial (or term).** Both words refer to a polynomial with exactly one piece. Ever polynomial can be thought of as a sum of monomials / terms. For example, the polynomial $x^7 - 4x^5 + 18x$ is the sum of the monomials x^7, $-4x^5$, and $18x$.
- **Degree (of a polynomial).** The largest power in a polynomial. The degrees of the polynomials $5x^3 - 2x + 1$, $-a^{99} - 9a^9 - 99$, and $4z^2 - 12z + 9$ are 3, 99, and 2. A constant is also a polynomial, and it has degree 0 if the constant itself is nonzero.
- **The zero polynomial.** The function $p(x) = \mathbf{0}$ is a polynomial as well, called the "zero polynomial". Note that this is not an equation. The polynomial always has value 0 no matter what x is. The zero polynomial has no terms and, strictly speaking, it has no degree either. Sometimes it is convenient to define the degree of the zero polynomial to be either -1 or negative infinity ($-\infty$).
- **Coefficient (of a term).** The number being multiplied by a power. The coefficient of x^3 in $5x^3 - 2x + 1$ is 5; the coefficient of a^9 in $-a^{99} - 9a^9 - 99$ is -9; the coefficient of z^7 in $4z^2 - 12z + 9$ is 0.
- **Function notation.** Often it is convenient to use function notation to represent long polynomials. When we say $f(x) = 5x^3 - 2x + 1$, we are saying that f "assigns" to a number x the value $5x^3 - 2x + 1$.
- **Zero of a polynomial** (also called **"root"** of a polynomial equation). The value

---

of x where the polynomial $P(x)$ has the value 0. It is a solution of the polynomial equation $P(x) = 0$.

- **Factored polynomial.** A polynomial is sometimes written as a product of different polynomials. For example, $(3x+1)(2x-7)$, or $(z^2 + 2z + 3)(-2z + 5)(7z^7 + 4z^4)$, or $(2y-8)^2(5y^2)^3$. This is sometimes done to make equations easier to solve, or to make values easier to compute.
- **Polynomial of multiple variables.** There can be multiple variables in a polynomial: $3xy^2$, $9x^5y^5 + 20x^3y^4$. The **degree** of these polynomials is the maximum sum of the exponents of variables. For example, $3xy^2$ has degree 3, and $9x^5y^5 + 20x^3y^4$ has degree 10.

Important Theorems

- **Fundamental Theorem of Algebra** (also known as the **Gauss-d'Alembert Theorem**.
    - ○ Every nonconstant polynomial with complex coefficients has at least one complex zero.
    - ○ Consequently, the number of zeros of a polynomial equals the degree, multiplicities counted.
    - ○ Note: The proof of this theorem is beyond the scope of this class. You can find more information about this theorem online.
- **Polynomial Remainder Theorem.** The remainder of a polynomial $P(x)$ divided by a linear divisor $x - a$ is equal to $P(a)$.

  Proof. Consider the polynomial division formula: Given polynomials $P(x)$ and $D(x)$, we can always write

  $$P(x) = D(x)Q(x) + R(x),$$

  where $\deg R(x) < \deg D(x)$. Here $Q(x)$ is the quotient, and $R(x)$ is the remainder. In the case of a linear divisor $D(x) = x - a$, the remainder is a constant r. So

  $$P(x) = (x-a)Q(x) + r.$$

  In the above identity, let $x = a$, then $P(a) = r$.   ∎

- **Factor Theorem.** A polynomial $P(x)$ has a factor $x - a$ if and only if $P(a) = 0$.

  Proof. This is just a special case (also called a "corollary") of the Polynomial Remainder Theorem when $r = 0$.   ∎

- **Vieta's Theorem (quadratic version).** Let x_1, x_2 be the roots of a quadratic equation $ax^2 + bx + c = 0$, where $a \neq 0$, then

  $$x_1 + x_2 = -\frac{b}{a}, \qquad \text{and} \qquad x_1 x_2 = \frac{c}{a}.$$

Proof. This is easily proven using the quadratic formula. Also see the proof of the general form below. ∎

- **Vieta's Theorem (general version).** Let $P(x) = a_n x^n + a_{n-1} x^{n-1} + \cdots + a_0$ be a polynomial of degree n, and $x_1, x_2, \ldots, x_n$ be the zeros of $P(x)$. Then

$$
\begin{cases}
x_1 + x_2 + \cdots + x_n &= -\dfrac{a_{n-1}}{a_n}, \\
x_1 x_2 + x_1 x_3 + \cdots + x_{n-1} x_n &= \dfrac{a_{n-2}}{a_n}, \\
\quad \cdots \\
x_1 x_2 \cdots x_n &= (-1)^n \dfrac{a_0}{a_n}.
\end{cases}
$$

Proof. By the Factor Theorem, since $x_1, x_2, \ldots, x_n$ are the zeros of $P(x)$, each of the linear polynomial $(x - x_1), (x - x_2), \ldots, (x - x_n)$ is a factor of $P(x)$. Thus

$$P(x) = a_n(x - x_1)(x - x_2) \cdots (x - x_n).$$

Expanding the right hand side, and compare the coefficients with $P(x) = a_n x^n + a_{n-1} x^{n-1} + \cdots + a_0$ to get the desired identities. ∎

5.1 Example Questions

Problem 5.1 True or False. If the statement is false, explain how to correct the statement.

(a) If the degree of a polynomial $P(x)$ is d, then the number of terms of $P(x)$ is between 1 and d (inclusive).

(b) If the degrees of polynomials $p(y)$ and $q(y)$ are d and e, then the degree of $p(y) \cdot q(y)$ is $d + e$.

(c) If the degrees of polynomials $N(y)$ and $M(y)$ are d and e, then the degree of $N(M(y))$ is e^d.

Problem 5.2 For this problem, $f(x) = 3x + 2$, $g(x) = x - 7$, and $h(x) = x^2 - 4x + 4$. Compute the following values:

(a) $f(g(4))$

(b) $g(g(g(g(g(35)))))$

(c) $h(f(0))$

(d) $h(f(100))$

(e) $f(g(1234567)) - g(f(1234567))$

Problem 5.3 In the polynomial $(7 + x)(1 + x^2)(5 + x^4)(2 + x^8)(3 + x^{16})(10 + x^{32})$, what is the coefficient of x^{54}?

Problem 5.4 Let $m \geq -1$ be a real number, and the equation $x^2 + 2(m - 2)x + m^2 - 3m + 3 = 0$ has two distinct real roots x_1 and x_2. If $x_1^2 + x_2^2 = 6$, what is m?

Problem 5.5 Assume $(x - c)^2 \mid (4x^3 + 8x^2 - 11x + 3)$, find the value of c.

Problem 5.6 Let a, b, c, and d be the roots of $x^4 - 2x - 1990 = 0$. Find the value of $1/a + 1/b + 1/c + 1/d$.

Problem 5.7 Expand $(x^2 - x + 1)^6$ to get $a_{12}x^{12} + a_{11}x^{11} + \cdots + a_1x + a_0$. Find the value of $a_{12} + a_{10} + a_8 + a_6 + a_4 + a_2 + a_0$.

Problem 5.8 Find the sum of the 17th powers of the 17 roots of $x^{17} - 3x + 1 = 0$.

Problem 5.9 An $l \times w \times h$ rectangular box has surface area 38 and volume 12. If $l + w + h = 8$, find the dimensions of the box.

Problem 5.10 Suppose that the roots of $3x^3 + 3x^2 + 4x - 11 = 0$ are a, b and c, and the roots of $x^3 + rx^2 + sx + t = 0$ are $a + b, b + c$, and $c + a$. Find t.

5.2 Quick Response Questions

Problem 5.11 Is a zero-degree polynomial the same as the zero polynomial?

Problem 5.12 Let $f(x) = 5x^3 - 2$ and $g(x) = x^2 + 2$. What is the degree of $f(g(x))$?

Problem 5.13 Let $f(x) = 5x^3 - 2$ and $g(x) = x^2 + 2$. What is the degree of $f(x) \times g(x)$?

Problem 5.14 What is the sum of the roots of $x^5 + 2x^4 + 3x^3 + 4x^2 + 5x + 6 = 0$?

Problem 5.15 Consider $x^3 - 6x^2 + 11x - 6 = 0$. Which of the following sets of roots are all POSSIBLE according to the Rational Root Theorem?

(A) $1, -4, 6$
(B) $0.5, 2, -6$
(C) $-1, 3, 6$
(D) $-2, -3, 12$

Problem 5.16 Find all the solutions to $x^3 - 3x^2 + 4x - 2$. Two of them are complex conjugates of the form $A \pm Bi$. What is $A + B$?

Problem 5.17 Find the sum of the 5th powers of the 5 roots of $x^5 + 2x - 1 = 0$.

Problem 5.18 Consider the complex roots of $x^4 - 16 = 0$. What is the product of the imaginary roots (the roots that are not real)?

Problem 5.19 How many real roots does $x^6 - 1 = 0$ have?

Problem 5.20 Suppose $P(x) = ax^3 + bx^2 + cx + d$. If we are given that $P(0) = 2$, $P(1) = 3$, and $P(-1) = 1$, what is b?

5.3 Practice Questions

Problem 5.21 True or False. If the statement is false, explain how to correct the statement.

(a) If the degree of every term in the polynomial $g(x)$ is even, and every coefficient is positive, then $g(x) \geq 0$ for every possible real value of x.

(b) If the degree of a single term in the polynomial $g(x)$ is odd, then $g(x) = 0$ for some real value of x.

Problem 5.22 For this problem, $f(x) = 3x + 2$, $g(x) = x - 7$, and $h(x) = x^2 - 4x + 4$. Compute the following values:

(a) $g(f(4))$

(b) $h(f(1))$

Problem 5.23 What is the coefficient of x^4 in $(x^4 + 3)(x^2 + 3x + 4)(x^2 - 2x - 3)$?

Problem 5.24 The two roots of equation $x^2 + px + 1 = 0$, where $p > 0$, have difference 1. Find the value of p.

Problem 5.25 Find all possible rational c such that $(x - c) \mid 2x^5 - 3x^4 - 2x^3 - 2x^2 + 3x + 2$.

Problem 5.26 If a, b, c, d are four different numbers for which

$$\begin{cases} a^4 + a^2 + ka + 64 &= 0 \\ b^4 + b^2 + kb + 64 &= 0 \\ c^4 + c^2 + kc + 64 &= 0 \\ d^4 + d^2 + kd + 64 &= 0. \end{cases}$$

What is the value of $a^2 + b^2 + c^2 + d^2$? Hint: Consider the polynomial $x^4 + x^2 + kx + 64$.

Problem 5.27 Assume $(3x - 1)^7 = a_7 x^7 + a_6 x^6 + \cdots + a_1 x + a_0$, find the value of $a_0 + a_2 + a_4 + a_6$.

Problem 5.28 Let x be a real number such that $x^3 + 4x = 8$. Determine the value of $x^7 + 64x^2$.

Problem 5.29 The volume of a box is 100 and the surface area is 160. Given that one of the sides is 2, what is the sum of all three dimensions? Try to use Viete's in your solution.

Problem 5.30 The polynomial $p(x) = x^3 + 2x^2 - 5x + 1$ has three different roots a, b, and c. Find $a^3 + b^3 + c^3$.

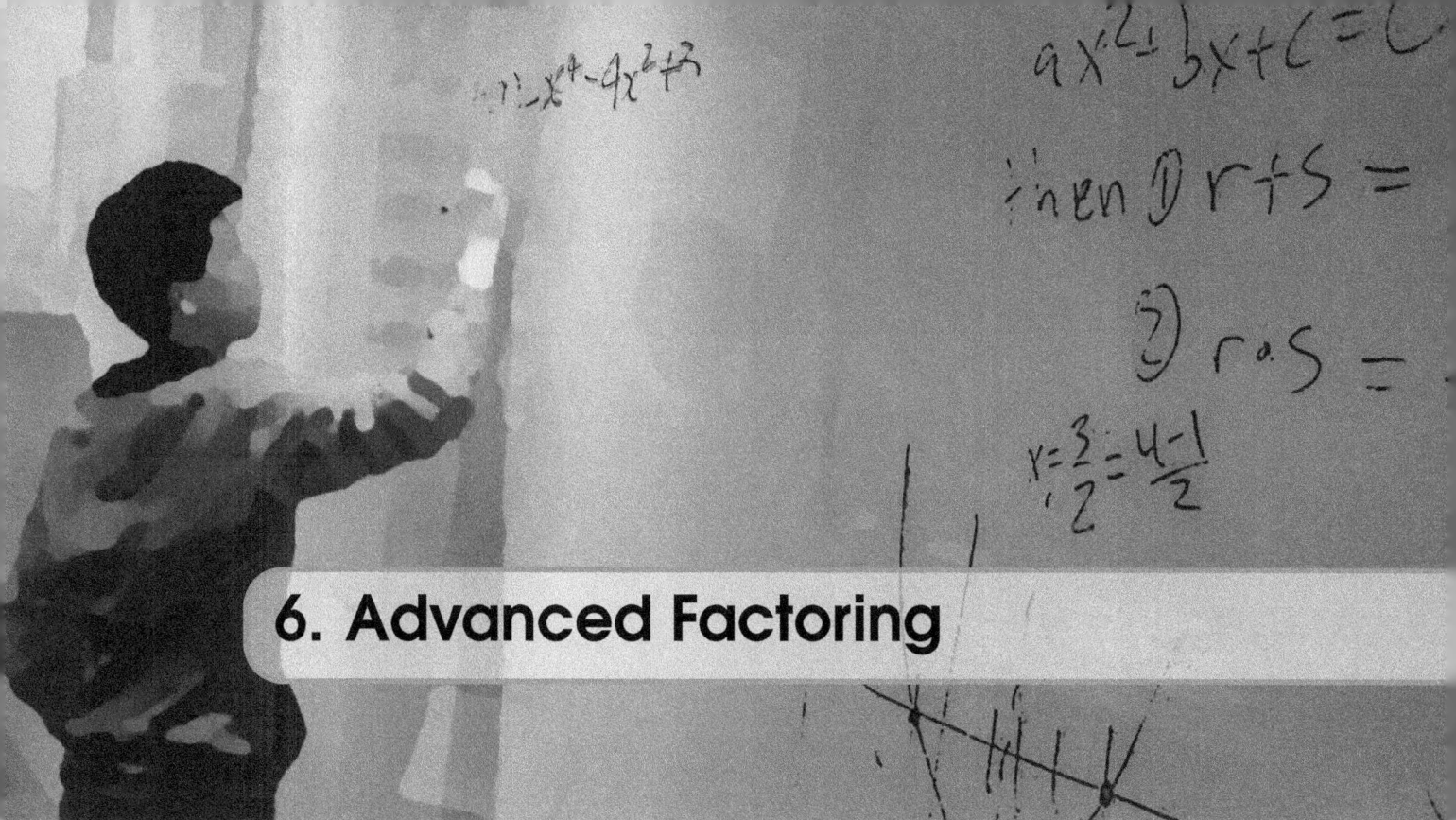

6. Advanced Factoring

Change of Variables

- A substitution or change of variables is often useful to replace a complicated part of the original expression, to make a simpler new expression.
- As a simple example, we can factor $x^4 + 2x^2 - 3$ as $(x^2 - 1)(x^2 + 3) = (x - 1)(x + 1)(x^2 + 3)$ using the substitution $x^2 = z$, so the equation becomes $z^2 + 2x - 3$ and can be factored as $(z - 1)(z + 3)$.

6.1 Example Questions

Problem 6.1 Factor the following.

(a) $(x^2 + x + 1)(x^2 + x + 2) - 12$. Hint: Try letting $y = x^2 + x + 1$.

(b) $(x^2 + 3x + 2)(x^2 + 7x + 12) - 120$. Hint: Factor and regroup so you can make the substitution $x^2 + 5x + 5$.

Problem 6.2 Factor the following using a change of variables.

(a) $x^2 + x - 14 - \dfrac{1}{x} + \dfrac{1}{x^2}$. Hint: Note $\left(x - \dfrac{1}{x}\right)^2 = x^2 - 2 + \dfrac{1}{x^2}$.

(b) $6x^4 + 7x^3 - 36x^2 - 7x + 6$.

Problem 6.3 Factor $(x+3)(x^2-1)(x+5) - 20$

Problem 6.4 Factor $(x^2 + xy + y^2)^2 - 4xy(x^2 + y^2)$. Hint: Let $u = x + y, v = xy$.

Problem 6.5 Factor $x^3 + 3x^2 - 4$

Problem 6.6 Factor $(x^2 + 4x + 8)^2 + 3x(x^2 + 4x + 8) + 2x^2$.

Problem 6.7 Factor $a^2 + (a+1)^2 + (a^2 + a)^2$

Problem 6.8 Factor the following.

(a) $2acx + 4bcx + adx + 2bdx + 4acy + 8bcy + 2ady + 4bdy$

(b) $1 + 2a + 3a^2 + 4a^3 + 5a^4 + 6a^5 + 5a^6 + 4a^7 + 3a^8 + 2a^9 + a^{10}$.

Problem 6.9 Factor $a^5 + a + 1$.

Problem 6.10 Evaluate the following: $\dfrac{(1994^2 - 2000)(1994^2 + 3985) \times 1995}{1991 \cdot 1993 \cdot 1995 \cdot 1997}$.

6.2 Quick Response Questions

Problem 6.11 Which of the following substitutions for z gives $(x+1)(x+3)+4 = z^2+2z+4$?

    (A) $z = x-1$
    (B) $z = x+1$
    (C) $z = x+3$
    (D) $z = x-3$

Problem 6.12 $8x^3 - 36x^2y + 54xy^2 - 27y^3$ can be factored as $(Sx+Ty)^3$. What is $S+T$?

Problem 6.13 The equation $x^6 - 3x^3 - 40 = 0$ has one (real) integer root. What is this root?

Problem 6.14 $x^4 - 10x^2 + 24$ can be fully factored as $(x-a)(x-b)(x^2-c)$. What is c?

Problem 6.15 Find the roots of $\dfrac{1}{x^2} - \dfrac{4}{x} + 3 = 0$. What is the sum of the roots? Round your answer to the nearest hundredth if necessary.

Problem 6.16 Factor $1 + a + a^2 + a^3$.

(A) $(a+1)(a^2+a+1)$
(B) $(a-1)(a^2+1)$
(C) $(a+1)(a^2+1)$
(D) $(a+1)(a-1)(a+1)$

Problem 6.17 Factor $(x^2+3x+1)(x^2+3x+4)+2$.

(A) $(x+1)(x+2)(x^2+3x+3)$
(B) $(x+1)(x+2)(x^2+3x+1)$
(C) $(x+1)(x+2)(x+3)(x+4)$
(D) $(x^2+3x-1)(x^2+3x+3)$

Problem 6.18 Factor $8 - \dfrac{36}{x} + \dfrac{54}{x^2} - \dfrac{27}{x^3}$.

(A) $(2/x-3)^3$
(B) $(3-2/x)^3$
(C) $(3/x-2)^3$
(D) $(2-3/x)^3$

Problem 6.19 Which of the following is equal to $x^3y + xy^3$ after the substitution $u = x+y$ and $v = xy$.

(A) $uv(u-2v)$
(B) $v(u^2-2v)$
(C) u^2+v
(D) uv^2

Problem 6.20 Write $\dfrac{2017^2 + 2017 + 1}{2017^3 - 1}$ as the reduced fraction $\dfrac{P}{Q}$ (with $P, Q \geq 1$ and $\gcd(P, Q) = 1$). What is $P + Q$?

6.3 Practice Questions

Problem 6.21 Factor $(x^2 + 3x + 2)(4x^2 + 8x + 3) - 90$.

Problem 6.22 Factor $x^4 + 7x^3 + 14x^2 + 7x + 1$

Problem 6.23 Factor $(x + 1)(x + 3)(x + 5)(x + 7) + 15$

Problem 6.24 Factor $(x + y)^4 - 6xy(x^2 + y^2) - 4x^2y^2$.

Problem 6.25 Factor $x^3 + 9x^2 + 26x + 24$

Problem 6.26 Factor $(2x^2 - 3x + 1)^2 - 22x^2 + 33x - 1$

Problem 6.27 Factor $ab(c^2 - d^2) - cd(a^2 - b^2)$

Problem 6.28 Factor $abc + 3b^2c + 3abd + 9b^2d + 3ac^2 + 9bc^2 + 9acd + 27bcd$.

Problem 6.29 Factor $a^5 + a^4 + 1$.

Problem 6.30 Evaluate

$$\frac{(2016^2 - 4032)(2016^2 + 2014)}{2014 \times 2016 \times 2018}.$$

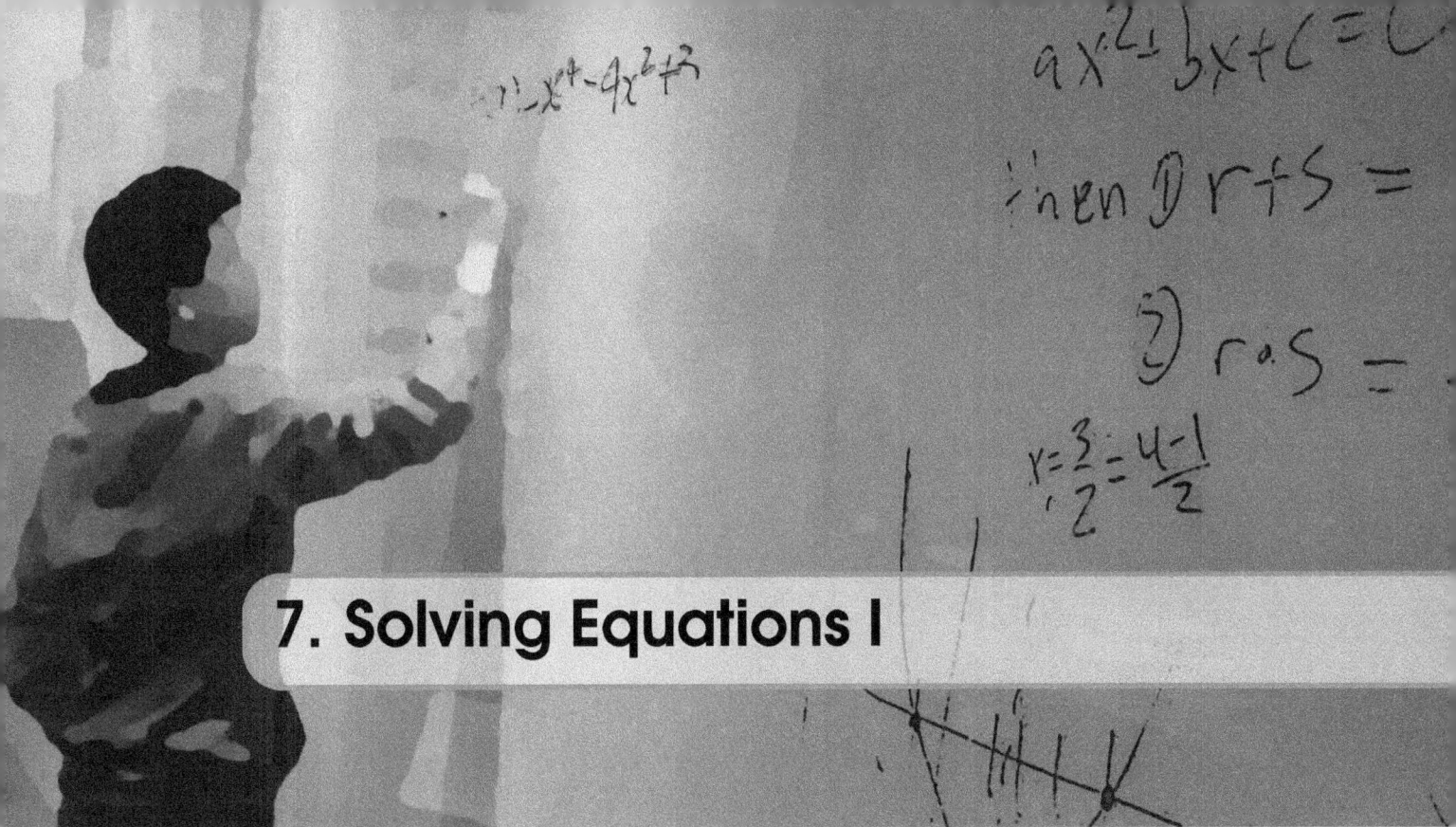

7. Solving Equations I

Solving Equations

- The standard method to solve equations is by factoring.
- Methods from previous weeks, such as change of variables are very helpful for factoring equations which can then lead to solutions.
- The goal of most substitutions is to turn the equation into a quadratic, which we can then solve by factoring or by using the quadratic formula.

Equations with Fractions

- When unknowns appear on denominators, the standard method is to multiply everything by the denominator (or the least common multiple of the denominators) to get rid of the denominators.
- This is likely to introduce extraneous roots. Therefore, it is always necessary to verify the roots at the end.
- As with other types of equations, a change of variables may be used to simplify the equations.

Equations with Absolute Values

- The absolute value of a real number a is defined as:

$$|a| = \begin{cases} a, & \text{if } a \geq 0; \\ -a, & \text{if } a < 0. \end{cases}$$

- The standard method to deal with absolute values is case analysis: solve in intervals where the expressions inside the absolute value do not change signs.
- Sometimes the following techniques also help:
    - Change of variables.
    - Using the property that absolute values are always nonnegative.

7.1 Example Questions

Problem 7.1 Solve the following:

(a) $\dfrac{15}{x+1} = \dfrac{15}{x} - \dfrac{1}{2}$.

(b) $\dfrac{4x}{x^2-4} - \dfrac{2}{x-2} = \dfrac{x+1}{x+2}$.

Problem 7.2 Solve the following:

(a) $\dfrac{3-x}{2+x} = 5 - \dfrac{4(2+x)}{3-x}$.

(b) $\dfrac{x-3}{x+1} - \dfrac{x+1}{3-x} = \dfrac{5}{2}$.

Problem 7.3 Solve the equation $\dfrac{1}{2x^2-3} - 8x^2 + 12 = 0$.

Problem 7.4 Solve the following equations over the reals by considering cases.

(a) $|x| + 2 = |2x|$.

(b) $|x^2 + 1| = 2|x - 1|$.

(c) $|x| - 2 = -|1 - x|$.

Problem 7.5 Solve the following.

(a) Solve: $|x - |2x + 1|| = 3$.

(b) Solve: $|x^2 - 11x + 10| = |2x^2 + x - 45|$.

Problem 7.6 If $|m - 2009| = -(n - 2010)^2$, what is $(m - n)^{2011}$?

Problem 7.7 The equation $|x^2 - 5x| = a$ has exactly two distinct real roots. What is the possible range of values for a?

Problem 7.8 Solve: $\left(\dfrac{x+1}{x^2-1}\right)^2 - 4\left(\dfrac{x+1}{x^2-1}\right) + 3 = 0$.

Problem 7.9 $(2x^2 - 3x + 1)^2 = 22x^2 - 33x + 1$

Problem 7.10 Solve $2x^4 - 9x^3 + 14x^2 - 9x + 2 = 0$.

7.2 Quick Response Questions

Problem 7.11 Find the sum of all solutions to $\dfrac{x+5}{x-5} = \dfrac{x+6}{x-6}$.

Problem 7.12 Find the product of all solutions to $\left(\dfrac{1}{y-1}\right)^2 + \dfrac{1}{y-1} = 6$. Round your answer to the nearest hundredth if necessary.

Problem 7.13 Solve $\dfrac{1}{x} + \dfrac{1}{x+2} = \dfrac{12}{5}$. If $r < s$ are the roots, what is $r - s$ rounded to the nearest tenth?

Problem 7.14 Solve $\dfrac{1}{x^2} - \dfrac{1}{x^2+3} = \dfrac{3}{4}$. What is the difference between the largest and smallest root?

Problem 7.15 Solve $|x+3| = |x|$. What is the smallest solution, rounded to the nearest hundredth if necessary?

Problem 7.16 How many integer solutions are there to the inequality $|3x| \le 9$?

Problem 7.17 Solve $|x^2 + 2| = 6$ over the real numbers. The solutions are $\pm K$ for an integer K. What is K?

Problem 7.18 Solve $|x + |x + 1|| = 4$. How many solutions are there?

Problem 7.19 Find all triples (x, y, z) such that $|x - 2| + |y^2 - 4| + |z - x| = 0$. For these triples, what is the largest possible value of $x + y + z$?

Problem 7.20 What is the smallest positive integer that is a solution to the inequality $|3x - 4| > 12$

7.3 Practice Questions

Problem 7.21 Solve (over the reals) $\dfrac{1}{x-1} + \dfrac{1}{x+1} = 3$.

Problem 7.22 Find the real solutions to $\dfrac{1}{x^2+2} - x^2 = \dfrac{1}{2}$.

Problem 7.23 Solve: $\dfrac{3x-1}{x^2+1} - \dfrac{3x^2+3}{3x-1} = 2$.

Problem 7.24 Find all the real solutions to $|x^2-1| = |x+1|$.

Problem 7.25 Find the real solutions of $|x^2-3x+1| = |x^2+x+2|$.

Problem 7.26 Solve the equation $|x^2-1| + |x^2+x-2| = 0$.

Problem 7.27 For what values of a, b the equation $x^2+2(1+a)x+(3a^2+4ab+4b^2+2) = 0$ has real roots?

Problem 7.28 Find all real solutions to $\dfrac{x+2}{x^2-4} + \dfrac{x+1}{x^2-x-2} = \dfrac{x}{x-2}$.

Problem 7.29 Find the real roots of $(x^2 + x + 1)(x^2 + x + 2) = 12$.

Problem 7.30 Find all real roots of $x^4 - 2x^3 - x^2 - 2x + 1 = 0$.

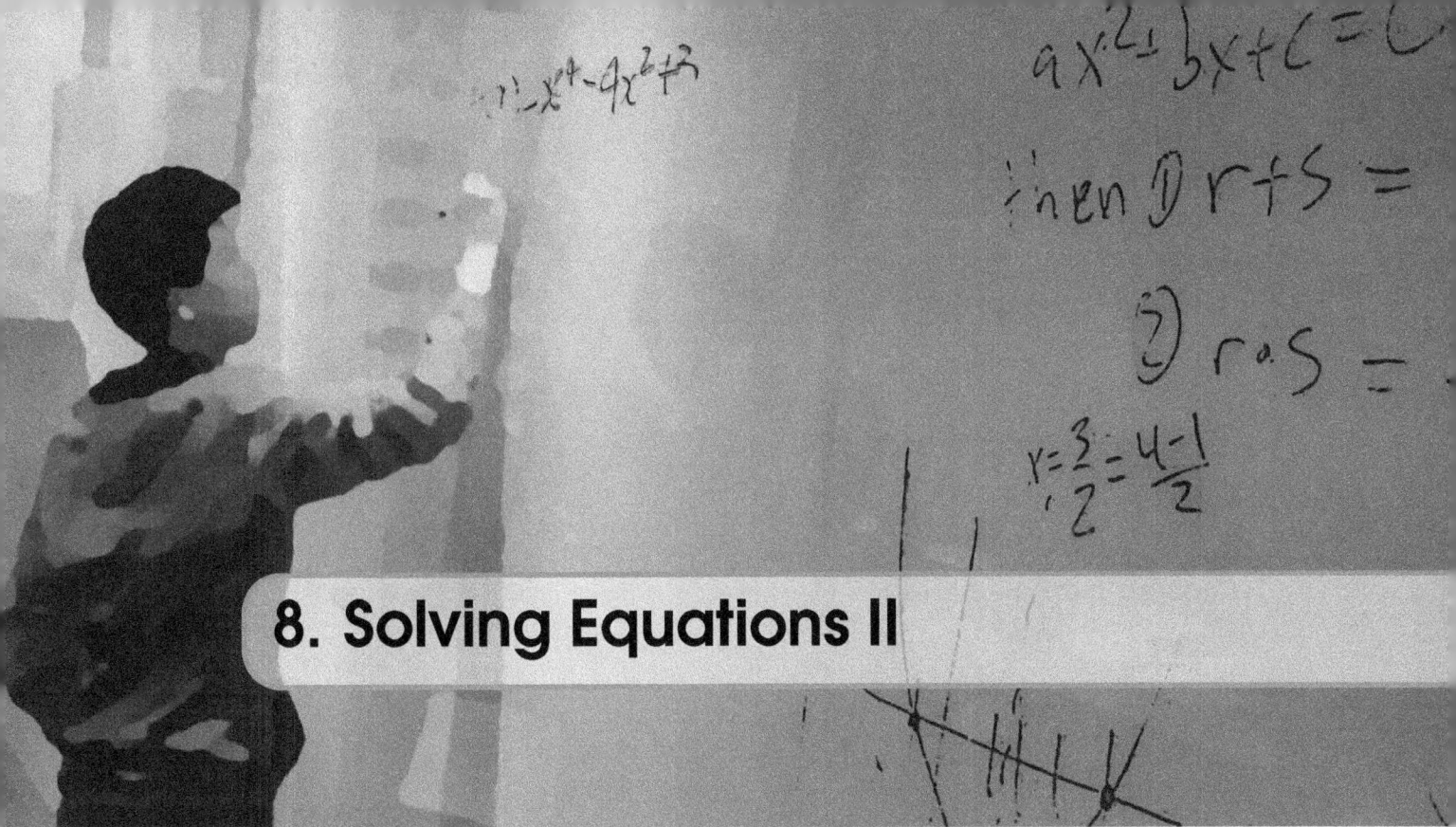

8. Solving Equations II

Equations with Radicals

- If unknown variables appear inside radicals, the common method is to square (or cube, etc., depending on the order of the roots) both sides to remove the radicals.
- Sometimes the following methods also help:
  - Change of variables
  - Using the fact that $\sqrt{a}$ is always nonnegative for $a \geq 0$.

8.1 Example Questions

Problem 8.1 Find the domain and range of the following functions.

(a) $y = \sqrt{x^2 + 3x - 4}$.

(b) Find the domain and range of $y = \sqrt{x^2 - 6x + 13}$.

Problem 8.2 Find the real solutions to the following.

(a) $3 - \sqrt{2x - 3} = x$.

(b) $\sqrt{x + 3} - \sqrt{3x - 2} = -1$.

Problem 8.3 Solve: $\sqrt{x^2 + 3x + 7} - \sqrt{x^2 + 3x - 9} = 2$.

Problem 8.4 Solve: $x^2 - \sqrt{3x^2 + 7} = 1$.

Problem 8.5 Solve: $\sqrt{\sqrt{x + 4} + 4} = x$

Problem 8.6 Solve for real x: $\sqrt{\dfrac{x - 2}{x + 2}} + \sqrt{\dfrac{9x + 18}{x - 2}} = 4$.

Problem 8.7 Solve for x: $(x - \sqrt{3})x(x + 1) + 3 - \sqrt{3} = 0$.

Problem 8.8 Let a be a real number, and the equation $x^2 + a^2x + a = 0$ has real roots for x. Find the maximum possible root x.

Problem 8.9 Solve $\sqrt{2x + 2} - \sqrt{x + 3} = \sqrt{x + 1} - \sqrt{2x + 4}$.

Problem 8.10 For what range of k does $\sqrt{2x^2 + 4} = x + k$ have real solutions?

8.2 Quick Response Questions

Problem 8.11 Solve $\sqrt{x^2-16}+|x-4|=0$. What is the largest solution?

Problem 8.12 Consider the equation $|x-4|=\sqrt{x}$. Which of the following quadratics has the same solutions?

(A) $x^2-8x+16=0$
(B) $x^2-7x+16=0$
(C) $x^2-9x+16=0$
(D) $x^2+9x-16=0$

Problem 8.13 The domain of $\sqrt{x+5}+\sqrt{7-x}$ can be written as the closed interval $[A,B]$. What is $A+B$?

Problem 8.14 The range of $y=\sqrt{x^2-16}+|x|$ is all y such that $y \geq K$ for an integer K. What is K?

Problem 8.15 How many solutions does $\sqrt{1+\sqrt{1+x}}=2$ have?

Problem 8.16 The equation $\sqrt{|x|+x}=8$ has one solution. What is it? Round your answer to the nearest tenth if necessary.

Problem 8.17 The equation $\sqrt{x+5}+\sqrt{x+7}=5$ has one solution of the form $\dfrac{P}{Q}$ for $P>1$ and $Q>1$ with $\gcd(P,Q)=1$. What is $P-Q$?

Problem 8.18 The equation $\dfrac{1}{\sqrt{x+2}} + \sqrt{x+2} = \dfrac{10}{3}$ has one solution of the form $\dfrac{P}{Q}$ for P an integer and $Q > 1$ with $\gcd(P,Q) = 1$. What is $P - Q$?

Problem 8.19 Solve the following equation for y:

$$-x^3 - x^2 y - 2x^2 + xy - 2x + y^2 - 4 = 0.$$

(A) $y = x^2 + 2$ or $y = -x - 2$
(B) $y = x^2 - 2$ or $y = x - 2$
(C) $y = x^2 + x + 2$ or $y = -x - 2$
(D) $y = -x^2 - x + 2$ or $y = x + 2$

Problem 8.20 Find the smallest real solution of $-x^3 - x^2 - 2x^2 + x - 2x + 1 - 4 = 0$.

8.3 Practice Questions

Problem 8.21 Find the domain and range of $y = \sqrt{81 - 9x^2} - \sqrt{9 - x^2}$.

Problem 8.22 Solve $\sqrt{x - 4} = x - 6$ for real values of x.

Problem 8.23 Solve $\sqrt{x^2 + 2x + 6} = \sqrt{x^2 + 6x + 3}$

Problem 8.24 Solve: $2x^2 - \sqrt{4x^2 - 12x} = 6x + 4$.

Problem 8.25 Solve for x: $\sqrt{5 - \sqrt{5 - x}} = x$.

Problem 8.26 Solve the equation $\sqrt{\dfrac{x}{x+2}} + \sqrt{\dfrac{x+2}{x}} = 2$.

Problem 8.27 Solve $16 + 4x - 4x^2 - x^3 = 0$.

Problem 8.28 Let a be a real number such that the equation $x^2 + a^2 x + a = 0$ has real roots for x. Find the maximum possible root x. For what value of a is the maximum root achieved?

Problem 8.29 Find the number of solutions to $\sqrt{x+2} - \sqrt{x+3} = \sqrt{x+1} - \sqrt{x+4}$.

Problem 8.30 The equation $2kx^2 + (8k+1)x + 8k = 0$ has two distinct real roots for x. Find the range of values for k.

9. Exponents and Logarithms

Inverses

- Given a function $f(x)$, we call $g(x)$ an *inverse* of $f(x)$ if $f(g(x)) = g(f(x)) = x$ for all x.
- Note not all functions have inverses. For example, the inverse of $y = x + 1$ is $y = x - 1$, but the function $y = x^2$ does not have an inverse.
- Sometimes it may be useful to restrict the domains. For example, the inverse of $y = x^2$ when $x \geq 0$ is $y = \sqrt{x}$.

Exponents

- Recall the rules for exponents:
  - $x^0 = 1$ when $x \neq 0$.
  - $x^a \cdot x^b = x^{a+b}$.
  - $\dfrac{x^a}{x^b} = x^{a-b}$.
  - $(x^a)^b = x^{ab}$.

Logarithms

- The inverse to the function $y = b^x$ (with $b > 0$) is the *logarithmic* function with base b, denoted $\log_b(x)$. Note the domain of the $\log_b(x)$ fuction is only positive numbers ($x > 0$).
- Therefore, if $b^x = y$ we have that $\log_b(y) = x$.

- Common bases are 10: $\log_{10}(x) = \log(x)$, e: $\log_e(x) = \ln(x)$ (the natural log), 2: $\log_2(x) = \lg(x)$.
- From the rules for exponents we get the following for logarithms.
    - $\log_b(1) = 0$.
    - $\log_b(x) + \log_b(y) = \log_b(x \cdot y)$.
    - $\log_b(x) - \log_b(y) = \log_b(x \div y)$.
    - $y \cdot \log_b(x) = \log_b(x^y)$.
- The "change of base" formula $\log_b(a) = \dfrac{\log_c(a)}{\log_c(b)}$ is often useful.

9.1   Example Questions

Problem 9.1 For each of the following statements, say whether they are true or false. If false, give a counterexample and correct the statement.

(a) Every linear function has an inverse.

(b) Every (non-constant) polynomial function $f(x)$ can be divided into pieces so that each piece has an inverse.

(c) If $f(g(x)) = x$, then $g(f(x)) = x$.

Problem 9.2 Find the inverse of the following functions.

(a) $f(x) = 4x + 3$.

(b) $f(x) = \dfrac{1}{x+1}$.

(c) $f(x) = x + \dfrac{1}{x}$ for $x \geq 1$. Note: $f(1) = 2$ is the minimum of $f(x)$.

Problem 9.3 Assume $x > 0$. Let $f(x) = x^2$, $g(x) = 2x + 3$, and $h(x) = \dfrac{1}{x}$. Calculate:

(a) $f^{-1}(g(h(2)))$.

(b) $h(g^{-1}(f^{-1}(36)))$

(c) $h(h(f(h(h(g(g^{-1}(g(x)))))))))$.

Problem 9.4 Prove $\log_b(x) + \log_b(y) = \log_b(x \cdot y)$ using the rules for exponents and the definition of logs as the inverse of the exponential function.

Problem 9.5 Calculate the following

(a) $\log_5(125)$.

(b) $\log_2(8) + \log_{\sqrt{2}}(4)$

(c) $\log(30) - \log(3)$.

(d) $\dfrac{\log_2(343)}{\log_2(49)}$.

Problem 9.6 What is the domain and range of $\log_3(\log_{1/3}(x+2))$? What are its zeros?

Problem 9.7 Solve the equation $\log_2(x) + \log_4(x) + \log_8(x) = \dfrac{22}{3}$.

Problem 9.8 For how many integer values is $\log(x-20) + \log(30-x) < 1$?

Problem 9.9 Solve $2\log_2(x)\log_4(x) + 2\log_2(x) - 3 = 0$.

Problem 9.10 (AIME 1983) Let x, y, and z all exceed 1, and let w be a positive number such that $\log_x(w) = 24$, $\log_y(w) = 40$, and $\log_{xyz}(w) = 12$. Find $\log_z(w)$.

9.2 Quick Response Questions

Problem 9.11 Calculate the inverse of $y = \dfrac{2x}{x-2}$.

(A) $\dfrac{2}{x-2}$

(B) $\dfrac{2x}{x-2}$

(C) $\dfrac{x}{x-2}$

(D) $\dfrac{2x}{x+2}$

Problem 9.12 Calculate the inverse of $y = x + \dfrac{2}{x}$ for $x \geq \sqrt{2}$.

(A) $\dfrac{1}{2}(x - \sqrt{x^2 - 8})$ for $x \geq \sqrt{2}$

(B) $\dfrac{1}{2}(x - \sqrt{x^2 - 8})$ for $x \geq 2\sqrt{2}$

(C) $\dfrac{1}{2}(x + \sqrt{x^2 - 8})$ for $x \geq \sqrt{2}$

(D) $\dfrac{1}{2}(x + \sqrt{x^2 - 8})$ for $x \geq 2\sqrt{2}$

Problem 9.13 Calculate $\log_2(256)$.

Problem 9.14 Calculate $\log_9(243)$. Round your answer to the nearest tenth if necessary.

Problem 9.15 $\log_8(256)$ can be written as $\dfrac{P}{Q}$ for $P, Q > 1$ with $\gcd(P, Q) = 1$. What is $P + Q$?

Problem 9.16 Is it true that $f(x) = \log(x^2)$ is the same as the function $g(x) = 2\log(x)$?

Problem 9.17 Calculate $\log(250) + \log(4)$. Round to the nearest integer if necessary.

Problem 9.18 Calculate $\log_4(80) - \log_4(5)$. Round to the nearest integer if necessary.

Problem 9.19 Solve $\log_x(2) + \log_2(x) = \dfrac{5}{2}$. What is the smaller root, rounded to the nearest tenth if necessary?

Problem 9.20 The four roots of the equation $y = \log_2(|x^2 - 4|)$ are $\pm\sqrt{K}$ and $\pm\sqrt{L}$ for integers $K \neq L$. What is $K + L$?

9.3 Practice Questions

Problem 9.21 Let $f(x)$ be the linear function $f(x) = mx + b$ with $m \neq 0$.

(a) Find a general formula for the inverse of $f(x)$.

(b) Describe all linear $f(x)$ so that $f^{-1}(x) = f(x)$.

Problem 9.22 Find the smallest m so that $f(x) = x^2 - 6x + 12$ for $x \geq m$ has an inverse. What is the inverse?

Problem 9.23 Let $f(x) = \dfrac{1}{x+1}, g(x) = x - 2, h(x) = \dfrac{1}{x}$.   Calculate $f(h(f(g(h^{-1}(10)))))$.

Problem 9.24 Prove the change of base formula $\log_b(a) = \dfrac{\log_c(a)}{\log_c(b)}$.

Problem 9.25 Write the following expressions as a single logarithm.

(a) $\log_2(x) + \log_4(x) + 2$.

(b) $\log_{\sqrt{3}}(x) + 2\log_3(x) - \log_9(x)$.

Problem 9.26 Find the domain of $\log_{100}(\log_{1/10}(\log_{10}(x^2-1)))$.

Problem 9.27 Solve the equation $\log_{\sqrt{3}}(x) + \log_3(x) + \log_9(x) = -\dfrac{21}{2}$.

Problem 9.28 Find the exact range of values so that $\log(x-20) + \log(30-x) < 1$.

Problem 9.29 Solve $2\log_{10}(x)\log_{10}(\sqrt{x}) = \log_{10}(100x)$.

Problem 9.30 Let $x, y, z > 1$ be such that $\log_{xy}(z) = 2$ and $\log_z(x/y) = 1$. What is $\log_y(z)$?

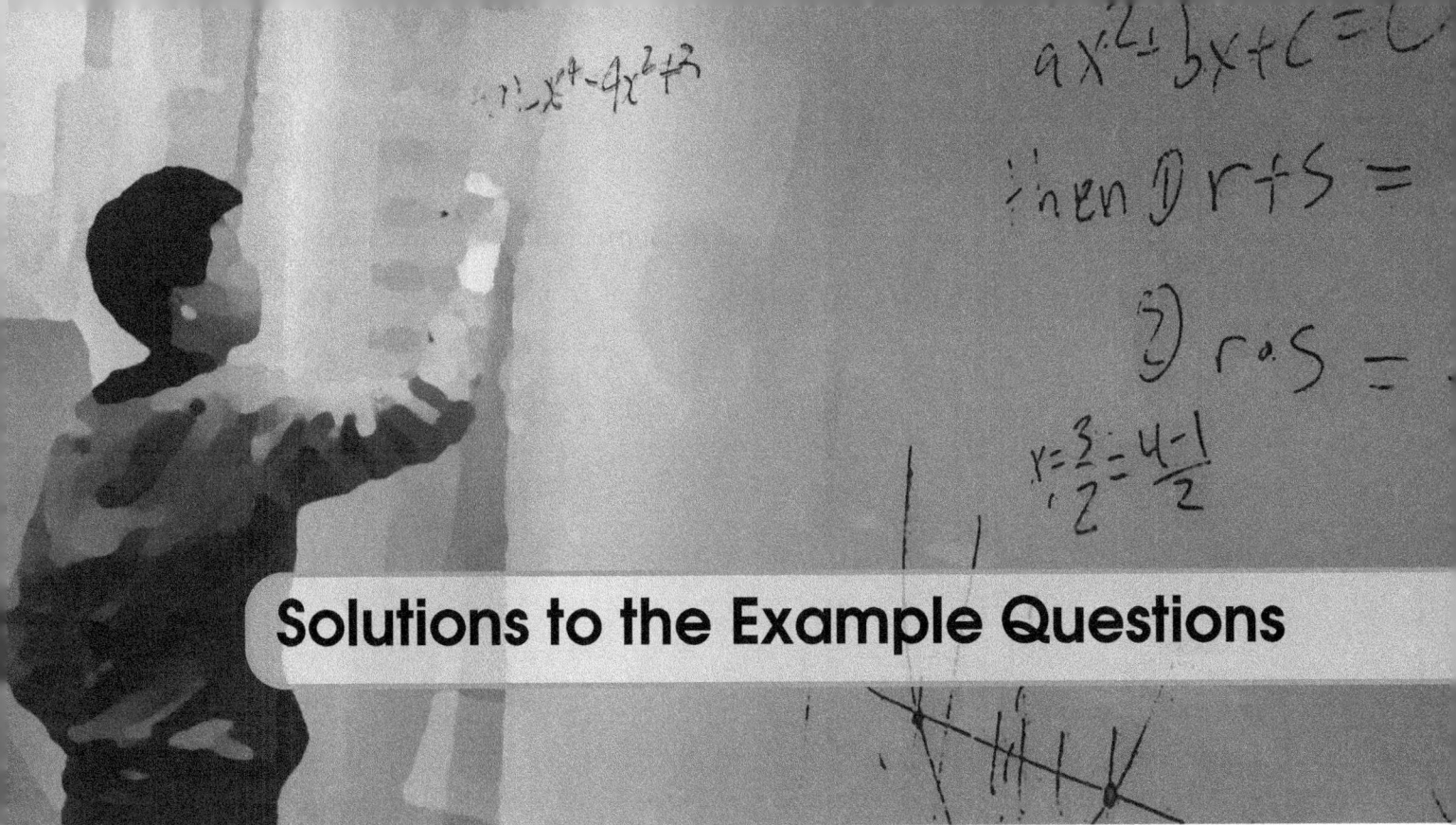

Solutions to the Example Questions

In the sections below you will find solutions to all of the Example Questions contained in this book.

Quick Response and Practice questions are meant to be used for homework, so their answers and solutions are not included. Teachers or math coaches may contact Areteem at info@areteem.org for answer keys and options for purchasing a Teachers' Edition of the course.

1 Solutions to Chapter 1 Examples

Problem 1.1 Factor the following.

(a) $10x^{10}y^8 + 5x^5y^9$.

> **Answer**

$5x^5y^8(2x^5 + y)$

> **Solution**

The constants share a common factor of 5, the x terms share a common factor of x^5, and the y terms share a common factor of y^8. Dividing the factor $5x^5y^8$ from both addends yields $2x^5 + y$.

(b) $(2x + 3y)^3 - 8x^3 - 27y^3$.

> **Answer**

$18xy(2x + 3y)$

> **Solution**

Expand $(2x + 3y)^3 = 8x^3 + 36x^2y + 54xy^2 + 27y^3$. Cancel like terms and factor $36x^2y + 54xy^2$. The constants share a common factor of 18, the x terms share a common factor of x and the y terms share a common factor of y. Dividing the factor $18xy$ from both addends yields $2x + 3y$.

Problem 1.2 Prove the following identities:

(a) Prove the formula for factoring $a^3 + b^3 + c^3 - 3abc$.

> **Solution**

$$
\begin{aligned}
a^3 + b^3 + c^3 - 3abc &= (a+b)^3 - 3ab(a+b) + c^3 - 3abc \\
&= (a+b)^3 + c^3 - 3ab(a+b+c) \\
&= (a+b+c)[(a+b)^2 - c(a+b) + c^2] - 3ab(a+b+c) \\
&= (a+b+c)(a^2 + b^2 + c^2 - ab - bc - ca)
\end{aligned}
$$

A second method to prove this formula would be to expand the factored form and see if the result is identical to the other side.

(b) Prove the formula for factoring $y^6 - y^3$.

Solution

$$\begin{aligned} y^6 - y^3 &= y^3(y^3 - 1) \\ &= y^3(y-1)(y^2 + y + 1) \end{aligned}$$

We could also do $y^6 - y^3 = (y^2)^3 - y^3$ and apply the difference of cubes formula. However, finding common factors first is usually simpler.

Problem 1.3 Factor $x^{15} + x^{14} + x^{13} + \cdots + x^2 + x + 1$.

Answer

$(x^8 + 1)(x^4 + 1)(x^2 + 1)(x + 1)$

Solution

Note that $(x - 1)(x^{15} + x^{14} + x^{13} + \cdots + x^2 + x + 1) = x^{16} - 1 = (x^8 + 1)(x^8 - 1) = (x^8 + 1)(x^4 + 1)(x^4 - 1) = (x^8 + 1)(x^4 + 1)(x^2 + 1)(x^2 - 1) = (x^8 + 1)(x^4 + 1)(x^2 + 1)(x + 1)(x - 1)$. Dividing both sides by $(x - 1)$ yields the desired result.

Problem 1.4 Factor the following.

(a) Factor $22y^2 - 35y + 3$.

Answer

$(2y - 3)(11y - 1)$

Solution

Use some trial and error to get $22 = 2 \times 11$ and $3 = (-3)(-1)$, and $-35 = 2(-1) + 11(-3)$, thus

$$22y^2 - 35y + 3 = (2y - 3)(11y - 1).$$

This factors the equation as needed.

(b) Factor $2x^2 - 7x + 3$.

Answer

$(2x-1)(x-3)$.

Solution

Note that the leading term's coefficient 2 can be factored into 2 and 1 and the constant term 3 can be factored into -3 and -1. Apply cross multiplication to determine that $2(-3)+1(-1)=-7$ is the middle term's coefficient. Therefore, $2x^2-7x+3=(2x-1)(x-3)$.

(c) Factor $3k^2-5k-2$.

Answer

$(3k+1)(k-2)$.

Solution

Note that the leading term's coefficient 3 can be factored into 3 and 1 and the constant term -2 can be factored into -2 and 1. Apply cross multiplication to determine that $3(-2)+1(1)=-5$ is the middle term's coefficient. Therefore, $3k^2-5k-2=(3k+1)(k-2)$.

Problem 1.5 Factor $a^4+a^3+a^2b+ab^2+b^3-b^4$.

Answer

$(a+b)(a^2+b^2)(a-b+1)$

Solution

$$
\begin{aligned}
& a^4+a^3+a^2b+ab^2+b^3-b^4 \\
= {}& (a^4-b^4)+(a^3+b^3)+(a^2b+ab^2) \\
= {}& (a+b)(a-b)(a^2+b^2)+(a+b)(a^2-ab+b^2)+(a+b)ab \\
= {}& (a+b)[(a^2+b^2)(a-b)+(a^2-ab+b^2)+ab] \\
= {}& (a+b)[(a^2+b^2)(a-b)+(a^2+b^2)] \\
= {}& (a+b)(a^2+b^2)(a-b+1).
\end{aligned}
$$

Problem 1.6 Factor x^3-9x+8 by using the given technique:

(a) $8 = -1 + 9$.

> **Solution**

$x^3 - 9x + 8 = x^3 - 9x - 1 + 9 = (x^3 - 1) - 9(x - 1) = (x - 1)(x^2 + x + 1) - 9(x - 1) = (x - 1)(x^2 + x + 1 - 9) = (x - 1)(x^2 + x - 8)$.

(b) $-9x = -x - 8x$.

> **Solution**

$x^3 - 9x + 8 = x^3 - x - 8x + 8 = (x^3 - x) - (8x - 8) = x(x + 1)(x - 1) - 8(x - 1) = (x^2 + x - 8)(x - 1)$.

(c) $x^3 = 9x^3 - 8x^3$.

> **Solution**

$x^3 - 9x + 8 = 9x^3 - 8x^3 - 9x + 8 = (9x^3 - 9x) - (8x^3 - 8) = 9x(x + 1)(x - 1) - 8(x - 1)(x^2 + x + 1) = (x - 1)(9x^2 + 9x - 8x^2 - 8x - 8) = (x - 1)(x^2 + x - 8)$.

(d) Add two terms: $-x^2 + x^2$.

> **Solution**

$x^3 - 9x + 8 = x^3 - x^2 + x^2 - 9x + 8 = (x^3 - x^2) + (x^2 - 9x + 8) = x^2(x - 1) + (x - 8)(x - 1) = (x^2 + x - 8)(x - 1)$.

Problem 1.7 Factor the following.

(a) $x^9 + x^6 + x^3 - 3$.

> **Answer**

$(x - 1)(x^2 + x + 1)(x^6 + 2x^3 + 3)$.

Solution

If we re-express $-3 = -1 - 1 - 1$, we have

$$
\begin{aligned}
x^9 + x^6 + x^3 - 3 &= (x^9 - 1) + (x^6 - 1) + (x^3 - 1) \\
&= (x^3 - 1)(x^6 + x^3 + 1) + (x^3 - 1)(x^3 + 1) + (x^3 - 1) \\
&= (x^3 - 1)[(x^6 + x^3 + 1) + (x^3 + 1) + 1] \\
&= (x - 1)(x^2 + x + 1)(x^6 + 2x^3 + 3).
\end{aligned}
$$

(b) (Sophie Germain's technique) $x^4 + 4y^4$.

Answer

$(x^2 + 2xy + 2y^2)(x^2 - 2x^2y^2 + 2y^2)$.

Solution

$x^4 + 4y^4 = x^4 + 4x^2y^2 + 4y^4 - 4x^2y^2 = (x^2 + 2y^2)^2 - (2xy)^2 = (x^2 + 2xy + 2y^2)(x^2 - 2x^2y^2 + 2y^2)$.

Problem 1.8 Factor the expression $(x+1)^4 + (x^2 - 1)^2 + (x-1)^4$.

Answer

$(3x^2 + 1)(x^2 + 3)$

Solution

If we add and subtract $(x-1)^2(x+1)^2$, we have

$$
\begin{aligned}
& (x+1)^4 + (x^2 - 1)^2 + (x-1)^4 \\
&= (x+1)^4 + 2(x+1)^2(x-1)^2 + (x-1)^4 - (x-1)^2(x+1)^2 \\
&= ((x+1)^2 + (x-1)^2)^2 - (x^2 - 1)^2 \\
&= (2x^2 + 2)^2 - (x^2 - 1)^2 \\
&= (2x^2 + 2 + x^2 - 1)(2x^2 + 2 - (x^2 - 1)) \\
&= (3x^2 + 1)(x^2 + 3).
\end{aligned}
$$

Problem 1.9 Factor $a^2 + b^2 + c^2 - 2bc + 2ca - 2ab$.

Answer

$(a-b+c)^2$

Solution

Recall that $(a+b+c)^2 = a^2+b^2+c^2+2ab+2bc+2ca$. Therefore, $(a+(-b)+c)^2 = a^2+(-b)^2+c^2+2a(-b)+2(-b)c+2ca = a^2+b^2+c^2-2ab-2bc+2ca$ and hence our expression factors as $(a-b+c)^2$.

Problem 1.10 Factor $-2x^{5n-1}y^n + 4x^{3n-1}y^{n+2} - 2x^{n-1}y^{n+4}$.

Answer

$-2x^{n-1}y^n(x^n-y)^2(x^n+y)^2$

Solution

The constants share a common factor of -2, the x terms share a common factor of x^{n-1}, and the y terms share a common factor of y^n. Dividing the factor $-2x^{n-1}y^n$ from both addends yields $x^{4n} - 2x^{2n}y^2 + y^4$. Note that, $x^{4n} - 2x^{2n}y^2 + y^4 = (x^{2n})^2 - 2x^{2n}y^2 + (y^2)^2 = (x^{2n}-y^2)^2$. Furthermore, note that, $x^{2n} - y^2 = (x^n)^2 - (y)^2 = (x^n-y)(x^n+y)$. Combining the identities yields the desired solution.

2   Solutions to Chapter 2 Examples

Problem 2.1 Answer the following using the discriminant.

(a) For what values of m does the equation $4x^2 + 8x + m = 0$ have two distinct real roots?

Answer

$m < 4$

Solution

$\Delta = 8^2 - 4 \cdot 4m = 64 - 16m > 0$, so $m < 4$.

(b) Given that the equation $x^2 - 2x - m = 0$ has no real roots, how many real roots does the equation $x^2 + 2mx + 1 + 2(m^2 - 1)(x^2 + 1) = 0$ have?

Answer

None.

Solution

Derive $m < -1$ from $\Delta_1 < 0$, and $\Delta_2 = -4(2m-1)(m+1)(2m+1)(2m-1) < 0$. Or show the left hand side is always positive.

Problem 2.2 Use Vieta's formula to solve the following.

(a) A quadratic equation has two roots $\dfrac{2}{3}$ and $-\dfrac{1}{2}$, what is this equation? (multiple answers are possible)

Answer

One example: $6x^2 - x - 2$

Solution

We have
$$\frac{2}{3} - \frac{1}{2} = \frac{1}{6} = -\frac{b}{a}, \frac{2}{3} \cdot -\frac{1}{2} = -\frac{1}{3} = \frac{c}{a}.$$

Thus the equation is $x^2 - \frac{1}{6}x - \frac{1}{3} = 0$, which can be re-written as $6x^2 - x - 2 = 0$. (Or any equation with coefficients proportional to this one.

(b) Two real numbers have sum -10 and product -5, find these two numbers.

Answer

$-5 + \sqrt{30}; -5 - \sqrt{30}$

Solution

These two numbers are the roots of the equation

$$x^2 + 10x - 5 = 0.$$

Solving to get $x = \dfrac{-10 \pm \sqrt{10^2 + 4(5)}}{2} = -5 \pm \sqrt{30}$. So the two numbers are $-5 + \sqrt{30}$ and $-5 - \sqrt{30}$.

(c) (2005 AMC 10B/12B) The quadratic equation $x^2 + mx + n = 0$ has roots that are twice those of $x^2 + px + m = 0$, and none of m, n, and p is zero. What is the value of n/p?

Answer

8.

Solution

Assume the two roots of the latter equation are x_1 and x_2, then the roots of the former equation are $2x_1$ and $2x_2$. By Viète' Theorem, $x_1 + x_2 = -p$, $x_1 x_2 = m$, and also $2x_1 + 2x_2 = -m$ and $(2x_1)(2x_2) = n$. So we get $m = 2p, 4m = n$. So $n/p = 8$.

Problem 2.3 Let x_1, x_2 be the two roots for equation $x^2 + x - 3 = 0$, find the value of $x_1^3 - 4x_2^2 + 19$.

Answer

0.

Solution

x_1, x_2 are roots of the equation $x^2 + x - 3$, so $x_1^2 + x_1 - 3 = 0$ and similarly $x_2^2 + x_2 - 3 = 0$.

Hence, $x_1^2 = 3 - x_1, x_2^2 = 3 - x_2$, and we have $x_1^3 - 4x_2^2 + 19 = x_1(3 - x_1) - 4(3 - x_2) + 19$. Simplifying and substituting again we have $3x_1 - x_1^2 + 4x_2 + 7 = 3x_1 - (3 - x_1) + 4x_2 + 7 = 4(x_1 + x_2) + 4$. By Viete's Theorem we have $x_1 + x_2 = -1$, so $4(x_1 + x_2) + 4 = 4(-1) + 4 = 0$.

Problem 2.4 Given the equation in x, $x^2 + 2mx + m + 2 = 0$:

(a) For what values of m does the equation have two (not necessarily distinct) positive roots?

Answer

$-2 < m \leq -1$

Solution

First note for the equation to have two real roots we have $\Delta = (2m)^2 - 4 \cdot (m + 2) = 4m^2 - 4m - 8 = 4(m - 2)(m + 1) \geq 0$, so $m \geq 2$ or $m \leq -1$.

If the roots are x_1, x_2, $x_1, x_2 > 0$ means that $x_1 + x_2 > 0$ and $x_1 \cdot x_2 > 0$. Thus by Vieta's Theorem we have (sum of roots) $-2m > 0$ and (product of roots) $m + 2 > 0$. Hence $m < 0$ and $m > -2$.

We therefore see that $m \geq 2$ is impossible, so $m \leq -1$, $m < 0$, and $m > -2$. hence $-2 < m \leq -1$.

(b) For what values of m does the equation have one positive root and one negative root?

Answer

$m < -2$

Solution

Similar to part (a), $\Delta \geq 0$ implies $m \geq 2$ or $m \leq -1$.

If the roots are x_1, x_2 with $x_1 < 0 < x_2$, we have $x_1 \cdot x_2 < 0$ (note we cannot say whether $x_1 + x_2$ is positive or negative). By Vieta's theorem, (product of roots) $m + 2 < 0$, so $m < -2$. Hence we must have $m \leq -1$ and $m < -2$, so $m < -2$.

Problem 2.5 (2005 AMC 10A/12A) There are two values of a for which the equation

$4x^2 + ax + 8x + 9 = 0$ has only one solution for x. What is the sum of those values of a?

Answer

-16

Solution

For the quadratic equation to have only one solution, the discriminant has to be 0. Thus $\Delta = (a+8)^2 - 4 \cdot 4 \cdot 9 = 0$. Therefore $(a+8)^2 = 144$, and then $a+8 = \pm 12$, which means $a = 4$ or $a = -20$. So the answer is $4 + (-20) = -16$.

Problem 2.6 Find k in each of the following scenarios:

(a) In the equation $x^2 - 402x + k = 0$, one of the roots plus three equals 80 times the other root.

Answer

$k = 1985$

Solution

The two roots are 397 and 5.

(b) Let x_1 and x_2 be the two roots of the equation $4x^2 - 8x + k = 0$. Suppose further that $\dfrac{1}{x_1} + \dfrac{1}{x_2} = \dfrac{8}{3}$.

Answer

$k = 3$

Solution

By Vieta's formulas, $x_1 + x_2 = 2, x_1 x_2 = \dfrac{k}{4}$. Thus

$$\frac{1}{x_1} + \frac{1}{x_2} = \frac{x_1 + x_2}{x_1 x_2} = \frac{8}{k} = \frac{8}{3}.$$

Therefore $k = 3$.

Problem 2.7 Do the following.

(a) The sum of squares of the roots of equation $x^2 + 2kx = 3$ is 10. Find the possible values of k.

Answer

± 1

Solution

Let x_1 and x_2 be the roots of the equation. Applying Vieta's Theorem onto the quadratic equation $x^2 + 2kx - 3 = 0$ yields $x_1 + x_2 = -2k$ and $x_1 x_2 = -3$. Note that, $4k^2 = (-2k)^2 = (x_1 + x_2)^2 = x_1^2 + 2x_1 x_2 + x_2^2 = 10 + 2(-3) = 4$. This implies that $k = \pm 1$.

(b) For equation $2x^2 + mx - 2m + 1 = 0$, the sum of squares of the two real roots is $\dfrac{29}{4}$. Find the value of m.

Answer

$m = 3$.

Solution

Use Vieta's Theorem, $x_1^2 + x_2^2 = (x_1 + x_2)^2 - 2x_1 x_2 = \left(\dfrac{m}{2}\right)^2 - (-2m + 1) = \dfrac{29}{4}$. Solve for m to get $m = 3$ and $m = -11$. The other solution $m = -11$ doesn't provide real roots for the equation, so throw away.

Problem 2.8 The two real roots of $x^2 + (m - 2)x + 5 - m = 0$ are both greater than 2. Find the possible range of values for real number m.

Answer

$-5 < m \leq -4$.

Solution

Note that $x^2 + (m - 2)x + 5 - m = (x - 2)^2 + (m + 2)(x - 2) + (m + 5)$. Since there exists two real roots, the discriminant must be nonnegative. Equivalently, $(m + 2)^2 - 4(m + 5) = m^2 + 4m + 4 - 4m - 20 = m^2 - 16 \geq 0$. Therefore $m \leq -4$. Applying Vieta's Theorem yields $(x_1 - 2)(x_2 - 2) = m + 5$. Given that the real roots must also be greater than 2, we have $(x_1 - 2)(x_2 - 2) > 0$. Therefore, $m + 5 > 0$ or $m > -5$.

Problem 2.9 If x_1 and x_2 are integer roots of equation $x^2 + mx + 2 - n = 0$, and $(x_1^2 + 1)(x_2^2 + 1) = 10$, how many possible pairs (m, n) are there?

Answer

6.

Solution

Since x_1 and x_2 are integers, the possible values for $x_1^2 + 1$ and $x_2^2 + 1$ are $1, 10$ and $2, 5$. In the first case, $x_1 = 0, x_2 = \pm 3$; and in the second case, $x_1 = \pm 1$ and $x_2 = \pm 2$. The order of x_1 and x_2 does not matter, since we only need to find the possible pairs of (m, n). By Viète's formulas, $-m = x_1 + x_2, 2 - n = x_1 x_2$, so there are 6 different pairs for (m, n).

Problem 2.10 Let x_1, x_2 be two positive integer roots of equation $x^2 + px + 1997 = 0$. Find the value of $\dfrac{p}{(x_1 + 1)(x_2 + 1)}$.

Answer

$-1/2$

Solution

Let x_1 and x_2 be the roots of the equation $x^2 + px + 1997 = 0$ that satisfies $x_1 > x_2 > 0$. By Vieta's Theorem, $x_1 x_2 = 1997$ and $x_1 + x_2 = -p$. Since 1997 is prime, $x_1 = 1997$ and $x_2 = 1$. Therefore, $p = -1998$ and $\dfrac{p}{(x_1 + 1)(x_2 + 1)} = \dfrac{-1998}{(1997 + 1)(1 + 1)} = -\dfrac{1}{2}$.

3   Solutions to Chapter 3 Examples

Problem 3.1 Calculation Practice

(a) $6 + 8i + \overline{4 + 3i}$.

Answer

$10 + 5i$.

Solution

$6 + 8i + \overline{4 + 3i} = 6 + 8i + 4 - 3i = 10 + 5i$.

(b) $(3 + 4i)(-2 + 3i)$

Answer

$-18 + i$.

Solution

$(3 + 4i)(-2 + 3i) = -6 - 8i + 9i + 12i^2 = -6 + i - 12 = -18 + i$.

(c) $\dfrac{1 + i}{2 - i}$.

Answer

$\dfrac{1}{5} + i\dfrac{3}{5}$.

Solution

$$\frac{1 + i}{2 - i} = \frac{(1 + i)(2 + i)}{(2 - i)(2 + i)} = \frac{2 + 2i + i + i^2}{4 + 1} = \frac{1}{5} + i\frac{3}{5}$$

(d) $(1 - i)^4$.

Answer

-4.

Solution

$(1-i)^4 = ((1-i)^2)^2 = (1-2i+i^2)^2 = (-2i)^2 = -4.$

Problem 3.2 Solve the following quadratics over the complex numbers.

(a) $x^2 - 4x + 8 = 0.$

Answer

$2 \pm 2i$

Solution

Using the quadratic formula we have the roots of $x^2 - 4x + 8 = 0$ are

$$\frac{-(-4) \pm \sqrt{(-4)^2 - 4 \cdot 8}}{2} = \frac{4 \pm \sqrt{-16}}{2} = 2 \pm 2i.$$

(b) $4x^2 - 12x + 29 = 0.$

Answer

$\frac{3}{2} \pm i\sqrt{5}$

Solution

Using the quadratic formula we have the roots of $4x^2 - 12x + 29 = 0$ are

$$\frac{-(-12) \pm \sqrt{(-12)^2 - 4 \cdot 4 \cdot 29}}{4 \cdot 2} = \frac{12 \pm \sqrt{-320}}{8} = \frac{3}{2} \pm i\sqrt{5}.$$

Problem 3.3 Prove the following, where z, w are complex numbers.

(a) $\overline{z \cdot w} = \overline{z} \cdot \overline{w}.$

Solution

$$
\begin{aligned}
\overline{z \cdot w} &= \overline{(\mathrm{Re}(z) + i\mathrm{Im}(z))(\mathrm{Re}(w) + i\mathrm{Im}(w))} \\
&= \overline{(\mathrm{Re}(w)\mathrm{Re}(z) - \mathrm{Im}(w)\mathrm{Im}(z)) + (\mathrm{Re}(w)\mathrm{Im}(z) + \mathrm{Im}(w)\mathrm{Re}(z))i} \\
&= (\mathrm{Re}(w)\mathrm{Re}(z) - \mathrm{Im}(w)\mathrm{Im}(z)) - (\mathrm{Re}(w)\mathrm{Im}(z) + \mathrm{Im}(w)\mathrm{Re}(z))i \\
&= (\mathrm{Re}(w) - \mathrm{Im}(w)i)(\mathrm{Re}(z) - \mathrm{Im}(z)i) \\
&= \overline{w} \cdot \overline{z}.
\end{aligned}
$$

(b) $z\overline{z} = |z|^2$.

Solution

$z\overline{z} = (\mathrm{Re}(z) + i\mathrm{Im}(z))(\mathrm{Re}(z) - i\mathrm{Im}(z)) = ((\mathrm{Re}(z))^2 + (\mathrm{Im}(z))^2) = |z|^2$.

(c) $\dfrac{z}{w} = \dfrac{z\overline{w}}{|w|^2}$.

Solution

$\dfrac{z}{w} = \dfrac{z\overline{w}}{w\overline{w}} = \dfrac{z\overline{w}}{|w|^2}$.

Problem 3.4 Convert between the rectangular and polar forms: write each of these numbers in the other forms.

(a) i

Answer

$i = \cos\frac{\pi}{2} + i\sin\frac{\pi}{2}$

Solution

Since $|i| = \sqrt{0^2 + 1^2} = 1$ and $\tan(\theta)$ is undefined, $\theta = \frac{\pi}{2}$ and $i = \cos\frac{\pi}{2} + i\sin\frac{\pi}{2}$.

(b) $\sqrt{3} + i$.

Answer

$\sqrt{3} + i = 2\left(\cos\frac{\pi}{6} + i\sin\frac{\pi}{6}\right)$

Solution

Since $|\sqrt{3}+i| = \sqrt{(\sqrt{3})^2 + 1^2} = 2$ and $\tan(\theta) = \frac{1}{\sqrt{3}}$, $\theta = \frac{\pi}{6}$ and $\sqrt{3}+i = 2(\cos\frac{\pi}{6} + i\sin\frac{\pi}{6})$.

(c) $2\cos(180°) + 2i\sin(180°)$.

Answer

-2

Solution

$2\cos(180°) + 2i\sin(180°) = 2(-1) + 0 = -2$

(d) $\cos(7\pi/6) + i\sin(7\pi/6)$.

Answer

$-\frac{\sqrt{3}}{2} - \frac{1}{2}i$

Solution

$\cos(7\pi/6) + i\sin(7\pi/6) = -\frac{\sqrt{3}}{2} - i/2$

Problem 3.5 What are the sets of points satisfying the following? Draw the diagrams.

(a) $|z| \leq 2$.

Answer

Circular region centered at the origin with radius 2, boundary included.

Solution

Note the modulus gives the "distance" to the origin.

(b) $\text{Re}(z) > \dfrac{1}{2}$.

Answer

The half plane whose x-coordinate is greater than $\dfrac{1}{2}$.

Solution

The real part of $z = x + iy$ is x, which is the x-coordinate when graphed in the complex plane.

(c) $\mathrm{Re}(z) = \mathrm{Im}(z)$.

Answer

The line $y = x$.

Solution

If $z = x + iy$ then $\mathrm{Re}(z) = x$ and $\mathrm{Im}(z) = y$ so we have the line $y = x$ when graphed.

(d) $\left| \dfrac{z-1}{z+1} \right| < 1$

Answer

This includes all points in the half plane with positive x-coordinates.

Solution

This means $|z - 1| < |z + 1|$, so the point represented by z is closer to 1 than to -1. Thus this is all points (x, y) with $x > 0$.

Problem 3.6 What is $\left(\dfrac{\sqrt{2}}{2}(-1 + i) \right)^{100}$?

Answer

-1.

Solution

$$\left(\frac{\sqrt{2}}{2}(-1 + i) \right)^{100} = \frac{2^{50}}{2^{100}}(-1 + i)^{100} = \frac{1}{2^{50}}(-2i)^{50} = (-i)^{50} = i^2 = -1.$$

Problem 3.7 Given that $2 + ai$ and $b + i$ are the two roots of the quadratic equation $x^2 + px + q = 0$ where p and q are real numbers. What are p and q?

Answer

$p = -4, q = 5$

Solution

By Vieta's formulas, $p = -(2 + ai + b + i) = -(b+2) - (a+1)i$, and $q = (2 + ai)(b + i) = (2b - a) + (2 + ab)i$. Since p and q are real numbers, $a + 1 = 0$ and $2 + ab = 0$. Thus $a = -1$, $b = 2$ and therefore $p = -4, q = 5$.

Problem 3.8 (2009 AMC 12) For what value of n is $i + 2i^2 + 3i^3 + \cdots + n \cdot i^n = 48 + 49i$?

Answer

97

Solution

Let E be the largest integer $\leq n$ and O be the largest integer $\leq n$. Then we know that the real part of the expression is

$$-2 + 4 - 6 + 8 - \cdots \pm E = 48.$$

Note by grouping we see that E must be 96, so $n = 96$ or $n = 97$. Similarly, using the imaginary part, we have
$$1 - 3 + 5 - \cdots \pm O = 49,$$
so $O = 97$ so $n = 97$ or 98. This implies that $n = 97$.

Problem 3.9 Prove the triangle inequality for complex numbers. That is, for complex numbers x and y, show that
$$|x + y| \leq |x| + |y|.$$

Solution

Note that $|x + y|^2 = (x + y)(\bar{x} + \bar{y})$ using identities proven before. Distributing $x\bar{x} + x\bar{y} + y\bar{x} + y\bar{y}$. Note that $x\bar{y} + y\bar{x} = 2 \cdot \text{Re}(xy)$ (double check this) so

$$|x + y|^2 = |x|^2 + 2\text{Re}(xy) + |y|^2 \leq |x|^2 + 2|x||y| + |y|^2 = (|x| + |y|)^2$$

which proves that $|x + y| \leq |x| + |y|$ as needed.

Problem 3.10 Suppose z is not real and $|z| = 1$, show that $w = \dfrac{z-1}{z+1}$ is a pure imaginary number.

Note that

$$w = \frac{z-1}{z+1} = \frac{(z-1)(\bar{z}+1)}{(z+1)(\bar{z}+1)} = \frac{|z|^2 + (z-\bar{z}) - 1}{|z|^2 + (z+\bar{z}) + 1} = \frac{1 + 2\mathrm{Im}(z)i - 1}{1 + 2\mathrm{Re}(z) + 1} = \frac{\mathrm{Im}(z)}{1 + \mathrm{Re}(z)}i.$$

Hence w is a pure imaginary number.

4 Solutions to Chapter 4 Examples

Problem 4.1 Review of Rectangular and Polar Form

(a) Convert the following to polar form: $4 - 4i$.

Answer

$4\sqrt{2}(\cos(7\pi/4) + i\sin(7\pi/4))$

Solution

First note the modulus is $\sqrt{4^2 + (-4)^2} = \sqrt{32} = 4\sqrt{2}$. Then note that $\tan(7\pi/4) = \dfrac{-4}{4} = -1$ (with $7\pi/4$ an angle in the fourth quadrant like $(4, -4)$). Hence in polar form we have $4\sqrt{2}(\cos(7\pi/4) + i\sin(7\pi/4))$.

(b) Convert the following to rectangular form: $4(\cos(\pi/3) + i\sin(\pi/3))$.

Answer

$2 + 2i\sqrt{3}$

Solution

We have

$$4(\cos(\pi/3) + i\sin(\pi/3)) = 4 \cdot \frac{1}{2} + 4 \cdot \frac{\sqrt{3}}{2} = 2 + 2i\sqrt{3}$$

in rectangular form.

Problem 4.2 Solve the following quadratics with complex coefficients. Then verify Viete's formulas still hold for the sum and product of the roots.

(a) $2x^2 + ix + 1 = 0$.

Answer

$x = \dfrac{1}{2}i, x = -i.$

Solution

We can factor $2x^2 + ix + 1 = (2x - i)(x + i) = 0$ so $x = i/2$ or $x = -i$. Note

$$\frac{i}{2} - i = \frac{-i}{2} \text{ and } \frac{i}{2} \cdot (-i) = \frac{1}{2}$$

as needed for Viete's formulas.

(b) $x^2 - 2x + i(3x - 6) = 0$.

Answer

$x = 2, x = -3i$.

Solution

Rewriting before we factor we get $x^2 + (3i - 2)x - 6i = (x - 2)(x + 3i) = 0$ so $x = 2$ or $x = -3i$. It is easy to see that Viete's formulas hold.

Problem 4.3 Perform the following multiplications using both rectangular and polar form to verify the result for multiplying complex numbers in polar form.

(a) $(1 + i) \cdot (1 - i)$.

Answer

$2 = 2(\cos(0) + i\sin(0))$.

Solution

In rectangular form, $(1 + i) \cdot (1 - i) = 1 - i^2 = 2$. In polar form we have $1 + i = \sqrt{2}(\cos(\pi/4) + i\sin(\pi/4))$ and $1 - i = \sqrt{2}(\cos(-\pi/4) + i\sin(-\pi/4))$ so

$$(1 + i)(1 - i) = \sqrt{2} \cdot \sqrt{2}(\cos(0) + i\sin(0)) = 2,$$

as needed.

(b) $\left(\frac{1}{2} + i\frac{\sqrt{3}}{2}\right) \cdot (2i)$.

Answer

$-\sqrt{3} + i = 2(\cos(5\pi/6) + i\sin(5\pi/6))$.

Solution

In rectangular form,

$$\left(\frac{1}{2} + i\frac{\sqrt{3}}{2}\right) \cdot (2i) = i - \sqrt{3}.$$

In polar form we have

$$\left(\frac{1}{2} + i\frac{\sqrt{3}}{2}\right) = \cos(\pi/3) + i\sin(\pi/3)$$

and $2i = 2(\cos(\pi/2) + i\sin(\pi/2))$ so

$$\left(\frac{1}{2} + i\frac{\sqrt{3}}{2}\right) \cdot (2i) = 1 \cdot 2(\cos(\pi/2 + \pi/3) + i\sin(\pi/2 + \pi/3))$$

which simplifies to $2(\cos(5\pi/6) + i\sin(5\pi/6)$.

Problem 4.4 Do the following:

(a) Write $(1 - i)^{40}$ as $a + ib$ with a and b real numbers and simplify your answer.

Answer

1048576

Solution

In polar form, $1 - i = \sqrt{2}(\cos(-45°) + i\sin(-45°))$. Hence

$$
\begin{aligned}
(1 - i)^{40} &= [\sqrt{2}(\cos(-45°) + i\sin(-45°))]^{40} \\
&= 2^{20}(\cos(40 \cdot (-45°)) + i\sin(40 \cdot (-45°))) \\
&= 2^{20} \cdot 1 \\
&= 1048576,
\end{aligned}
$$

as $40 \cdot 45$ is a multiple of 360.

(b) Write $(\sqrt{3} + i)^8$ in polar and in rectangular form.

Answer

$2^8(\cos(4\pi/3) + i\sin(4\pi/3) = -128 - 128i\sqrt{3}$

Solution

$\sqrt{3}+i=2(\cos(\pi/6)+i\sin(\cos(\pi/6)))$. Hence

$$(\sqrt{3}+i)^8=[2(\cos(\pi/6)+i\sin(\cos(\pi/6)))]^8=2^8(\cos(8\pi/6)+i\sin(8\pi/6)).$$

Simplifying we have $2^8(\cos(4\pi/3)+i\sin(4\pi/3))=-128-128i\sqrt{3}$.

Problem 4.5 Roots of Unity Practice

(a) Write the 4th roots of unity in rectangular form.

Answer

$1,i,-1,-i.$

Solution

This is routine from the formula.

(b) Write (in rectangular form) the 12th roots of unity with argument (strictly) between $\dfrac{\pi}{2}$ and π.

Answer

$$-\frac{1}{2}+i\frac{\sqrt{3}}{2},-\frac{\sqrt{3}}{2}+i\frac{1}{2}$$

Solution

The 12th roots of unity are $\cos(k\cdot\pi/6)+i\sin(k\cdot\pi/6)$ for $k=0,1,2,\dots,11$. $k\pi/6$ is between $\pi/2$ and π for $k=4,5$ giving roots of unity

$$\cos(4\pi/6)+i\sin(4\pi/6)=-\frac{1}{2}+i\frac{\sqrt{3}}{2}$$

and

$$\cos(5\pi/6)+i\sin(5\pi/6)=-\frac{\sqrt{3}}{2}+i\frac{1}{2}.$$

Problem 4.6 Solve each of the following equations (over the complex numbers) and graph the solutions in the complex plane. If the answer cannot be simplified, it is okay to leave it in polar form.

(a) $x^5 = -32$.

> **Answer**

$-2\left(\cos\dfrac{2k\pi}{5} + i\sin\dfrac{2k\pi}{5}\right)$, for $k = 0, 1, 2, 3, 4$

> **Solution**

The solutions are -2 times the 5th roots of unity.

(b) $x^2 = i$.

> **Answer**

$\pm\left(\cos\dfrac{\pi}{4} + i\sin\dfrac{\pi}{4}\right) = \pm\left(\dfrac{\sqrt{2}}{2} + \dfrac{\sqrt{2}}{2}i\right)$

> **Solution**

If $x^2 = i$ then $x^4 = -1$ so $x^8 = 1$. Hence the solutions are 2 of the 8th roots of unity.

(c) $x^3 = 8i$.

> **Answer**

$-2i, \sqrt{3} + i, -\sqrt{3} + i$.

> **Solution**

Note $(-2i)^3 = 8i$, so the solutions are $-2i$ times the 3rd roots of unity.

Problem 4.7 If $|a| = |b| = 1$ and $a + b + 1 = 0$, what are a and b?

> **Answer**

$-\dfrac{1}{2} \pm \dfrac{\sqrt{3}}{2}i$.

> **Solution**

Let $a = \cos\alpha + i\sin\alpha$ and $b = \cos\beta + i\sin\beta$, then $\cos\alpha + i\sin\alpha + \cos\beta + i\sin\beta + 1 = 0$. Here we get $\cos\alpha + \cos\beta + 1 = 0$ and $\sin\alpha + \sin\beta = 0$. Hence, (1) $\alpha = 2k\pi - \beta$ or

(2) $\alpha = (2k+1)\pi + \beta$. In case (1), $\cos \alpha = \cos \beta$, thus they are both $-\dfrac{1}{2}$, and then we have $a = -\dfrac{1}{2} \pm \dfrac{\sqrt{3}}{2}i$, and $b = -a$. In case (2), $\cos \alpha = -\cos \beta$ and there is no solution.

Conclusion: a and b are $-\dfrac{1}{2} \pm \dfrac{\sqrt{3}}{2}i$.

Problem 4.8 (2002 AMC 12A) Find the number of ordered pairs of real numbers (a,b) such that $(a+bi)^{2002} = a - bi$. Note: You should also be able to describe all such solutions.

Answer

2004

Solution

Let $z = a + bi$. Then we want $z^{2002} = \bar{z}$. As $|z| = |\bar{z}|$ we must have either $|z| = 0$ or $|z| = 1$.

If $|z| = 0$ we have $z = 0$ (so $a = b = 0$), resulting in one pair.

If $|z| = 1$ we have

$$z^{2002} = \bar{z} \Rightarrow z^{2003} = z \cdot \bar{z} = |z|^2 = 1$$

and hence $z^{2003} = 1$. Hence the 2003rd roots of unity are all solutions, giving 2003 additional pairs.

In total we have 2004 solutions/pairs.

Problem 4.9 Find all the roots (including complex) to $x^4 + x^2 + 1 = 0$. Write your answer in rectangular form.

Answer

$\pm \dfrac{1}{2} \pm \dfrac{\sqrt{3}}{2}$

Solution

Note multiplying by $x^2 - 1$ we have

$$(x^4 + x^2 + 1)(x^2 - 1) = x^6 - 1$$

Hence all the solutions to $x^4 + x^2 + 1 = 0$ will be 6th roots of unity. However, $x = \pm 1$ are clearly not solutions to $x^4 + x^2 + 1 = 0$, so the only solutions are $\cos(k\pi/3) + i\sin(k\pi/3)$ for $k = 1, 2, 4, 5$ giving $\pm\dfrac{1}{2} \pm \dfrac{\sqrt{3}}{2}$ when written in rectangular form.

Problem 4.10 Let $\omega_1 = \cos\dfrac{2\pi}{n} + i\sin\dfrac{2\pi}{n}$ (so ω_1 is an nth root of unity). If a is an nth root of unity, prove that all n solutions to $z^n = 1$ are $a, a \cdot \omega_1, a \cdot (\omega_1)^2, \ldots, a \cdot (\omega_1)^{n-1}$.

Solution

First note $(\omega_1)^0, (\omega_1)^1, (\omega_1)^2, \ldots, (\omega_1)^{n-1}$ is equal to $\cos\dfrac{k \cdot 2\pi}{n} + i\sin\dfrac{k \cdot 2\pi}{n}$ for $k = 0, 1, 2, \ldots, n-1$. Further, as a is an nth root of unity, there is an integer j such that $a = \cos\dfrac{j \cdot 2\pi}{n} + i\sin\dfrac{j \cdot 2\pi}{n}$. Hence $a, a \cdot (\omega_1)^2, \ldots, a \cdot (\omega_1)^{n-1}$ is

$$\left(\cos\frac{j \cdot 2\pi}{n} + i\sin\frac{j \cdot 2\pi}{n}\right) \cdot \left(\cos\frac{k \cdot 2\pi}{n} + i\sin\frac{k \cdot 2\pi}{n}\right)$$
$$= \cos\frac{(j+k) \cdot 2\pi}{n} + i\sin\frac{(j+k) \cdot 2\pi}{n}$$

for $k = 0, 1, \ldots, n-1$. Since $\cos(2\pi + k) = \cos k$ and $\sin(2\pi + k) = \sin k$ we see this is equal to (possibly rearranged)

$$\cos\frac{l \cdot 2\pi}{n} + i\sin\frac{l \cdot 2\pi}{n}$$

for $l = 0, 1, \ldots, n-1$, the nth roots of unity as needed.

Note it is very useful to drawn your own diagram to help visualize this problem.

5 Solutions to Chapter 5 Examples

Problem 5.1 True or False. If the statement is false, explain how to correct the statement.

(a) If the degree of a polynomial $P(x)$ is d, then the number of terms of $P(x)$ is between 1 and d (inclusive).

Answer

False.

Solution

The number of terms is between 1 and $d + 1$. For example, polynomials $2a^3$, $3x^3 - 2x$, $7t^3 + 5t^2 + 2t$, and $100u^3 + 4u^2 - 3u - 20$ are all polynomials of degree 3.

(b) If the degrees of polynomials $p(y)$ and $q(y)$ are d and e, then the degree of $p(y) \cdot q(y)$ is $d + e$.

Answer

True.

Solution

For example, $p(y) = x^2 + 2$ (degree 2) and $q(y) = x^3 - x$ (degree 3), then $p(y)q(y) = x^5 + x^3 - 2x$ (degree 5).

(c) If the degrees of polynomials $N(y)$ and $M(y)$ are d and e, then the degree of $N(M(y))$ is e^d.

Answer

False.

Solution

The degree is $d \cdot e$. For example, if $N(y) = x^2 + 2$ (degree 2) and $M(y) = x^3 - x$ (degree 3), then $N(M(y)) = (x^3 - x)^2 + 2 = x^6 - 2x^4 + x^2 + 2$ (degree 6).

Problem 5.2 For this problem, $f(x) = 3x + 2$, $g(x) = x - 7$, and $h(x) = x^2 - 4x + 4$. Compute the following values:

(a) $f(g(4))$

Answer

-7.

(b) $g(g(g(g(g(35)))))$

Answer

0.

(c) $h(f(0))$

Answer

0.

(d) $h(f(100))$

Answer

90000.

(e) $f(g(1234567)) - g(f(1234567))$

Answer

-14.

Problem 5.3 In the polynomial $(7+x)(1+x^2)(5+x^4)(2+x^8)(3+x^{16})(10+x^{32})$, what is the coefficient of x^{54}?

Answer

14.

Solution

Converted to binary, $54_{10} = 110110_2$. Thus the power $x^{54} = x^{32} \cdot x^{16} \cdot x^4 \cdot x^2$, and the coefficient is the product of the constant terms whose power of x does not appear, so the answer is $7 \cdot 2 = 14$.

Problem 5.4 Let $m \geq -1$ be a real number, and the equation $x^2 + 2(m-2)x + m^2 - 3m + 3 = 0$ has two distinct real roots x_1 and x_2. If $x_1^2 + x_2^2 = 6$, what is m?

Answer

$$m = \frac{5 - \sqrt{17}}{2}.$$

Solution

There are two distinct real roots, so $\Delta = 4(m-2)^2 - 4(m^2 - 3m + 3) = -4m + 4 > 0$. Therefore $m < 1$, and thus $-1 \leq m < 1$. So $6 = x_1^2 + x_2^2 = (x_1 + x_2)^2 - 2x_1x_2 = 4(m-2)^2 - 2(m^2 - 3m + 3) = 2m^2 - 10m + 10$. Solve for m we get $m = \frac{5 \pm \sqrt{17}}{2}$. Given the $-1 \leq m < 1$ range, we get $m = \frac{5 - \sqrt{17}}{2}$.

Problem 5.5 Assume $(x-c)^2 \mid (4x^3 + 8x^2 - 11x + 3)$, find the value of c.

Answer

$1/2$.

Solution

Use the Rational Root Theorem to factor

$$4x^3 + 8x^2 - 11x + 3 = (2x-1)^2(x+3) = 4\left(x - \frac{1}{2}\right)^2(x+3).$$

Problem 5.6 Let a, b, c, and d be the roots of $x^4 - 2x - 1990 = 0$. Find the value of $1/a + 1/b + 1/c + 1/d$.

Answer

$-1/995$.

Solution

$abc + bcd + cda + dab = 2$, and $abcd = -1990$, so $1/a + 1/b + 1/c + 1/d = (abc + bcd + cda + dab)/abcd = -1/995$. Another way to solve this: Let $y = 1/x$, then $1 - 2y^3 - 1990y^4 = 0$, and the four roots for y are $1/a, 1/b, 1/c, 1/d$, so the answer is $-(-2)/(-1990) = -1/995$.

Problem 5.7 Expand $(x^2 - x + 1)^6$ to get $a_{12}x^{12} + a_{11}x^{11} + \cdots + a_1x + a_0$. Find the value of $a_{12} + a_{10} + a_8 + a_6 + a_4 + a_2 + a_0$.

Answer

365.

Solution

Let $x = 1$:
$$1^6 = a_{12} + a_{11} + a_{10} + \cdots + a_2 + a_1 + a_0.$$

Let $x = -1$:
$$3^6 = a_{12} - a_{11} + a_{10} - \cdots + a_2 - a_1 + a_0.$$

Adding,
$$730 = 2(a_{12} + a_{10} + a_8 + a_6 + a_4 + a_2 + a_0),$$

So $a_{12} + a_{10} + a_8 + a_6 + a_4 + a_2 + a_0 = 365$.

Problem 5.8 Find the sum of the 17th powers of the 17 roots of $x^{17} - 3x + 1 = 0$.

Answer

-17.

Solution

Each of the roots satisfy $x_i^{17} - 3x_i + 1 = 0$, thus $x_i^{17} = 3x_i - 1$. Using Vieta's formula, $\sum x_i^{17} = \sum(3x_i - 1) = 3\sum x_i - 17 = -17$.

Problem 5.9 An $l \times w \times h$ rectangular box has surface area 38 and volume 12. If $l + w + h = 8$, find the dimensions of the box.

Answer

$4 \times 3 \times 1$.

Solution

Since $lwh = 12$ and $2(lw + wh + hl) = 38$, the values l, w, h are the three roots of polynomial equation $x^3 - 8x^2 + 19x - 12 = 0$. Factor the polynomial, $(x-1)(x-3)(x-4) = 0$, so the three roots are $1, 3, 4$. Thus it is a $4 \times 3 \times 1$ box.

Problem 5.10 Suppose that the roots of $3x^3 + 3x^2 + 4x - 11 = 0$ are a, b and c, and the roots of $x^3 + rx^2 + sx + t = 0$ are $a+b, b+c$, and $c+a$. Find t.

Answer

5.

Solution

$t = -(a+b)(b+c)(c+a) = -(ab+bc+ca)(a+b+c)+abc = -(4/3)(-1)+(11/3) = 5$. Note: It is somewhat hard to come up with this factoring, but keep in mind we want terms such as $abc, ab+bc+ac, a+b+c$ to appear in the factoring.

6 Solutions to Chapter 6 Examples

Problem 6.1 Factor the following.

(a) $(x^2 + x + 1)(x^2 + x + 2) - 12$. Hint: Try letting $y = x^2 + x + 1$.

Answer

$(x - 1)(x + 2)(x^2 + x + 5)$.

Solution

Let $y = x^2 + x + 1$, the expression becomes $y(y + 1) - 12 = y^2 + y - 12 = (y + 4)(y - 3)$, then change back to x.

(b) $(x^2 + 3x + 2)(x^2 + 7x + 12) - 120$. Hint: Factor and regroup so you can make the substitution $x^2 + 5x + 5$.

Answer

$(x^2 + 5x + 16)(x + 6)(x - 1)$.

Solution

$$
\begin{aligned}
(x^2 + 3x + 2)(x^2 + 7x + 12) - 120 &= (x + 1)(x + 2)(x + 3)(x + 4) - 120 \\
&= (x^2 + 5x + 4)(x^2 + 5x + 6) - 120.
\end{aligned}
$$

Let $y = x^2 + 5x + 5$, then $(y - 1)(y + 1) - 120 - y^2 - 121 = (y + 11)(y - 11)$. We can then change back to x.

Problem 6.2 Factor the following using a change of variables.

(a) $x^2 + x - 14 - \dfrac{1}{x} + \dfrac{1}{x^2}$. Hint: Note $\left(x - \dfrac{1}{x}\right)^2 = x^2 - 2 + \dfrac{1}{x^2}$.

Answer

$\left(x - \dfrac{1}{x} - 3\right)\left(x - \dfrac{1}{x} + 4\right)$.

Solution

We an rewrite the equation as

$$x^2 - 2 + \frac{1}{x^2} + x - \frac{1}{x} - 12,$$

so after the subsitution $y = x - \frac{1}{x}$ we have $y^2 + y - 12 = (y+4)(y-3)$. Rewriting in terms of x gives the final answer.

(b) $6x^4 + 7x^3 - 36x^2 - 7x + 6$.

Answer

$(2x+1)(x-2)(3x-1)(x+3)$

Solution

$6x^4 + 7x^3 - 36x^2 - 7x + 6 = x^2\left(6x^2 + 7x - 36 - \frac{7}{x} + \frac{6}{x^2}\right)$. Let $y = x - \frac{1}{x}$, $y^2 = x^2 - 2 + \frac{1}{x^2}$, thus

$6x^4 + 7x^3 - 36x^2 - 7x + 6 = x^2(6(y^2 + 2) + 7y - 36) = x^2(2y - 3)(3y + 8) = (2xy - 3x)(3xy + 8x) = (2x^2 - 3x - 2)(3x^2 + 8x - 3) = (2x+1)(x-2)(3x-1)(x+3)$

Problem 6.3 Factor $(x+3)(x^2 - 1)(x+5) - 20$

Answer

$(x^2 + 4x - 7)(x^2 + 4x + 5)$

Solution

We have $(x+3)(x^2 - 1)(x+5) - 20 = (x+3)(x+1)(x-1)(x+5) - 20 = (x^2 + 4x + 3)(x^2 + 4x - 5) - 20$, then let $y = x^2 + 4x$. We thus have $(y+3)(y-5) - 20 = y^2 - 2y - 35 = (y-7)(y+5)$. Substituting back x gives the final answer $(x^2 + 4x - 7)(x^2 + 4x + 5)$.

Problem 6.4 Factor $(x^2 + xy + y^2)^2 - 4xy(x^2 + y^2)$. Hint: Let $u = x+y, v = xy$.

Answer

$(x^2 - xy + y^2)^2$

Solution

Making the subsitution, note $x^2 + y^2 = (x+y)^2 - 2xy = u^2 - 2v$. Thus the equation becomes

$$(u^2 - v)^2 - 4v(u^2 - 2v) = u^4 - 2u^2v + v^2 - 4u^2v + 8v^2 = u^4 - 6u^2v + 9v^2 = (u^2 - 3v)^2.$$

Substituting back in for u, v we have $(x^2 + 2xy + y^2 - 3xy)^2 = (x^2 - xy + y^2)^2$.

Problem 6.5 Factor $x^3 + 3x^2 - 4$

Answer

$(x-1)(x+2)^2$

Solution

Note we can split up $-4 = -1 - 3$ so we have $x^3 - 1 + 3(x^2 - 1)$. Factoring gives

$$\begin{aligned}
(x-1)(x^2 + x + 1) + 3(x-1)(x+1) &= (x-1)(x^2 + x + 1 + 3x + 3) \\
&= (x-1)(x^2 + 4x + 4) \\
&= (x-1)(x+2)^2
\end{aligned}$$

as our final answer.

Problem 6.6 Factor $(x^2 + 4x + 8)^2 + 3x(x^2 + 4x + 8) + 2x^2$.

Answer

$(x+2)(x+4)(x^2 + 5x + 8)$

Solution

Let $z = x^2 + 4x + 8$. Then our equation becomes $z^2 + 3xz + 2x^2$. Thinking of this as a quadratic in z, we can factor: $z^2 + 3xz + 2x^2 = (z + 2x)(z + x)$. Resubstituting we have

$$\begin{aligned}
(x^2 + 4x + 8 + 2x)(x^2 + 4x + 8 + x) &= (x^2 + 6x + 8)(x^2 + 5x + 8) \\
&= (x+2)(x+4)(x^2 + 5x + 8)
\end{aligned}$$

as our expression fully factored.

Problem 6.7 Factor $a^2 + (a+1)^2 + (a^2 + a)^2$

Answer

$(a^2 + a + 1)^2$

Solution

Expand the first two terms, $a^2 + (a+1)^2 = a^2 + a^2 + 2a + 1 = 2(a^2 + a) + 1$, so $a^2 + (a+1)^2 + (a^2 + a)^2 = (a^2 + a)^2 + 2(a^2 + a) + 1 = (a^2 + a + 1)^2$.

Problem 6.8 Factor the following.

(a) $2acx + 4bcx + adx + 2bdx + 4acy + 8bcy + 2ady + 4bdy$

Answer

(a+2b)(2c+d)(x+2y)

Solution

Grouping,

$$
\begin{aligned}
& (2acx + 4bcx) + (adx + 2bdx) + (4acy + 8bcy) + (2ady + 4bdy) \\
= {} & 2cx(a + 2b) + dx(a + 2b) + 4cy(a + 2b) + 2dy(a + 2b) \\
= {} & (a + 2b)(2cx + dx + 4cy + 2dy) \\
= {} & (a + 2b)((2c + d)x + (2c + d)(2y)) \\
= {} & (a + 2b)(2c + d)(x + 2y)
\end{aligned}
$$

(b) $1 + 2a + 3a^2 + 4a^3 + 5a^4 + 6a^5 + 5a^6 + 4a^7 + 3a^8 + 2a^9 + a^{10}$.

Answer

$(a + 1)^2(a^2 + a + 1)^2(a^2 - a + 1)^2$.

Solution

Grouping, this equals $(1 + a + a^2 + a^3 + a^4 + a^5)^2$, so answer is $(a + 1)^2(a^2 + a + 1)^2(a^2 - a + 1)^2$.

Problem 6.9 Factor $a^5 + a + 1$.

Answer

$(a^2 + a + 1)(a^3 - a^2 + 1)$

Solution

Add and minus a^2:

$$a^5 - a^2 + a^2 + a + 1 = a^2(a^3 - 1) + (a^2 + a + 1) = (a^2 + a + 1)(a^3 - a^2 + 1).$$

Problem 6.10 Evaluate the following: $\dfrac{(1994^2 - 2000)(1994^2 + 3985) \times 1995}{1991 \cdot 1993 \cdot 1995 \cdot 1997}$.

Answer

1996.

Solution

Let $x = 1994$, then the expression is

$$
\begin{aligned}
\frac{(x^2 - x - 6)(x^2 + 2x - 3)(x + 1)}{(x - 3)(x - 1)(x + 1)(x + 3)} &= \frac{(x + 2)(x - 3)(x + 3)(x - 1)(x + 1)}{(x - 3)(x - 1)(x + 1)(x + 3)} \\
&= x + 2 \\
&= 1996.
\end{aligned}
$$

7   Solutions to Chapter 7 Examples

Problem 7.1 Solve the following:

(a) $\dfrac{15}{x+1} = \dfrac{15}{x} - \dfrac{1}{2}$.

Answer

$x = -6$ and $x = 5$.

Solution

Multiplying $2x(x+1)$ we get $30x = 30(x+1) - (x(x+1))$ so $30x = 30x + 30 - x^2 - x$ or $x^2 + x - 30 = 0$. Therefore $(x+6)(x-5) = 0$ so $x = -6$ and $x = 5$ are solutions (double check neither are extraneous).

(b) $\dfrac{4x}{x^2-4} - \dfrac{2}{x-2} = \dfrac{x+1}{x+2}$.

Answer

$x = 1$

Solution

Multiplying $(x+2)(x-2)$ to get $4x - 2(x+2) = (x+1)(x-2)$, solve and get roots 1 and 2. 2 is extraneous, so $x = 1$.

Problem 7.2 Solve the following:

(a) $\dfrac{3-x}{2+x} = 5 - \dfrac{4(2+x)}{3-x}$.

Answer

$x = 1/2$ and $x = -1$.

Solution

Let $y = \dfrac{3-x}{2+x}$, solve and get $x = 1/2$ and $x = -1$.

(b) $\dfrac{x-3}{x+1} - \dfrac{x+1}{3-x} = \dfrac{5}{2}$.

Answer

$x = 7$ and $x = -5$.

Solution

Let $y = \dfrac{x-3}{x+1}$. Then $x = 7$ and $x = -5$. Both are roots after verifying.

Problem 7.3 Solve the equation $\dfrac{1}{2x^2 - 3} - 8x^2 + 12 = 0$.

Answer

$x = \pm\dfrac{\sqrt{7}}{2}$ and $x = \pm\dfrac{\sqrt{5}}{2}$.

Solution

Let $y = 2x^2 - 3$, then $\dfrac{1}{y} - 4y = 0$. Solve and get $y = \pm\dfrac{1}{2}$, and solve for x, get $x = \pm\dfrac{\sqrt{7}}{2}$ and $x = \pm\dfrac{\sqrt{5}}{2}$, all are verified to be solutions.

Problem 7.4 Solve the following equations over the reals by considering cases.

(a) $|x| + 2 = |2x|$.

Answer

$x = \pm 2$.

Solution

Consider cases based on whether x is negative or not. If $x \geq 0$, then we have $x + 2 = 2x$ so $x = 2$. If $x < 0$, then we have $-x + 2 = -2x$ so $x = -2$. These are the two solutions.

(b) $|x^2 + 1| = 2|x - 1|$.

Answer

$x = -1 \pm \sqrt{2}$.

Solution

Note $|x^2+1| = x^2+1$ because squares are always positive. If $x \geq 1$ we have $x^2+1 = 2x-2$ so $x^2-2x+3 = 0$. Note the discriminant is negative, so this has no real roots. If $x < 1$ we have $x^2+1 = 2-2x$. Hence $x^2+2x-1 = 0$ and we can solve for x to get $x = -1 \pm \sqrt{2}$ using the quadratic formula.

(c) $|x| - 2 = -|1-x|$.

Answer

$-\dfrac{1}{2}, \dfrac{3}{2}$.

Solution

Consider three cases: (i) $x < 0$, (ii) $0 \leq x < 1$, or (iii) $x \geq 1$. In case (i) we have $-x-2 = -1+x$ so $2x = -1$ and $x = -1/2$. In case (ii), $x-2 = -1+x$ so $-2 = -1$ which is impossible. Lastly, for (iii), $x-2 = -x+1$ so $2x = 3$ and $x = 3/2$.

Problem 7.5 Solve the following.

(a) Solve: $|x - |2x+1|| = 3$.

Answer

$x = 2$ or $x = -4/3$

Solution

Case work: $x \geq -\dfrac{1}{2}$ or $x < -\dfrac{1}{2}$. If $x \geq -\frac{1}{2}$, $|x+1| = 3$, so $x = 2$ or $x = -4$ (throw away -4). If $x < -\dfrac{1}{2}$, $|3x+1| = 3$, then $x = 2/3$ (throw away) or $x = -4/3$. Final solution: $x = 2$ or $x = -4/3$.

(b) Solve: $|x^2 - 11x + 10| = |2x^2 + x - 45|$.

Answer

$x = -6 \pm \sqrt{91}$ and $x = \dfrac{5 \pm \sqrt{130}}{3}$.

Solution

It is a bit tedious to do case work on $x < 1$, $1 \leq x < 9/2$, $9/2 \leq x < 5$, $5 \leq x < 10$ and $x \geq 10$. The faster way is to solve $x^2 - 11x + 10 = 2x^2 + x - 45$ and $x^2 - 11x + 10 = -(2x^2 + x - 45)$, and get $x = -6 \pm \sqrt{91}$ and $x = \dfrac{5 \pm \sqrt{130}}{3}$.

Problem 7.6 If $|m - 2009| = -(n - 2010)^2$, what is $(m - n)^{2011}$?

Answer

-1.

Solution

No squares are negative, and no absolute values are negative. So $m = 2009, n = 2010$, and $m - n = -1$. So $(m - n)^{2011} = -1$.

Problem 7.7 The equation $|x^2 - 5x| = a$ has exactly two distinct real roots. What is the possible range of values for a?

Answer

$a = 0$ or $a > 25/4$.

Solution

$a = 0$ is obviously a good value. If $a > 0$, the two equations $x^2 - 5x - a = 0$ and $x^2 - 5x + a = 0$ combined have two distinct roots, that means the second equation has no real roots. Therefore, $5^2 + 4a > 0$ and $5^2 - 4a < 0$, which lead to $a > -\dfrac{25}{4}$ and $a > \dfrac{25}{4}$, thus $a > \dfrac{25}{4}$.

Problem 7.8 Solve: $\left(\dfrac{x+1}{x^2-1}\right)^2 - 4\left(\dfrac{x+1}{x^2-1}\right) + 3 = 0$.

Answer

$x = 2$ and $x = 4/3$.

$x = -1$ is not a root, so cancel it and simplify. Then let $y = \dfrac{1}{x-1}$. Solve and get $x = 2$ and $x = 4/3$. Both are roots after verifying.

Problem 7.9 $(2x^2 - 3x + 1)^2 = 22x^2 - 33x + 1$

Answer

$-3/2, 0, 3/2, 3$.

Solution

Let $y = (2x^2 - 3x + 1)$. This gives $y^2 = 11y - 10$ so $y^2 - 11y + 10 = 0$ and $(y - 10)(y - 1) = 0$ so $y = 10, y = 1$. Setting $10, 1$ equal to $2x^2 - 3x + 1$ gives $x = -3/2, 0, 3/2, 3$ as our solutions.

Problem 7.10 Solve $2x^4 - 9x^3 + 14x^2 - 9x + 2 = 0$.

Answer

$x = 1$, $x = 2$, and $x = 1/2$.

Solution

Divide everything by x^2: $2\left(x^2 + \dfrac{1}{x^2}\right) - 9\left(x + \dfrac{1}{x}\right) + 14 = 0$. Let $y = x + \dfrac{1}{x}$, then $y^2 = x^2 + \dfrac{1}{x^2} + 2$, so $2(y^2 - 2) - 9y + 14 = 0$, and $y = 2$ and $y = 5/2$. So $x = 1$, $x = 2$, and $x = 1/2$.

8   Solutions to Chapter 8 Examples

Problem 8.1 Find the domain and range of the following functions.

(a) $y = \sqrt{x^2 + 3x - 4}$.

Answer

Domain: $x \le -4$ or $x \ge 1$. Range: $y \ge 0$.

Solution

$x^2 + 3x - 4 = (x + 4)(x - 1)$ so this is non-negative when $x \le -4$ or $x \ge 1$. This gives the domain as we can take the square root of any non-negative number. Since $x^2 + 3x - 4$ can be any non-negative number, $y = \sqrt{x^2 + 3x - 4}$ can be any $y \ge 0$.

(b) Find the domain and range of $y = \sqrt{x^2 - 6x + 13}$.

Answer

Domain: All real numbers. Range: $y \ge 2$.

Solution

Completing the square we have $x^2 - 6x + 13 = x^2 - 6x + 9 + 4 = (x - 3)^2 + 4$ so we have $y = \sqrt{(x-3)^2 + 4}$. Hence the function is always defined. Further, $(x - 3)^2 \ge 0$, so $y \ge \sqrt{4} = 2$.

Problem 8.2 Find the real solutions to the following.

(a) $3 - \sqrt{2x - 3} = x$.

Answer

$x = 2$

Solution

$3 - x = \sqrt{2x - 3}$, so $9 - 6x + x^2 = 2x - 3$, and $x^2 - 8x + 12 = 0$, then $x = 2$ and $x = 6$. Check and find that $x = 2$ is the solution.

(b) $\sqrt{x + 3} - \sqrt{3x - 2} = -1$.

Answer

$x = 6$.

Solution

$\sqrt{x+3} = \sqrt{3x+2} - 1$, squaring, $x+3 = 3x-2-2\sqrt{3x-2}+1$, thus $x-2 = \sqrt{3x-2}$, square again, $x^2 - 7x + 6 = 0$, and get $x = 1$ or 6. $x = 1$ is extraneous. Therefore $x = 6$.

Problem 8.3 Solve: $\sqrt{x^2+3x+7} - \sqrt{x^2+3x-9} = 2$.

Answer

$x = 3$ and $x = -6$.

Solution

Let $y = \sqrt{x^2+3x+7}$, then $y - \sqrt{y^2-16} = 2$, then $y^2 - 4y + 4 = y^2 - 16$, so $y = 5$. Solve for x to get $x = 3$ and $x = -6$, both check out alright.

Problem 8.4 Solve: $x^2 - \sqrt{3x^2+7} = 1$.

Answer

$x = \pm\sqrt{6}$.

Solution

Multiply 3 on both sides. Let $y = \sqrt{3x^2+7}$. Then $y = 5$, so $x = \pm\sqrt{6}$.

Problem 8.5 Solve: $\sqrt{\sqrt{x+4}+4} = x$

Answer

$x = \dfrac{\sqrt{17}+1}{2}$.

Solution

First note that $x > 0$. Square both sides:

$$\sqrt{x+4}+4 = x^2.$$

Then $\sqrt{x+4} = x^2 - 4$, also note that $x > 2$.

Squaring again to get

$$x + 4 = x^4 - 2 \cdot 4x^2 + 4^2.$$

To avoid factoring a quartic polynomial in x, we apply the following technique: who says x must be the variable and 4 has to be the constant? Treat the constant 4 as a variable, and x as a parameter. Then it is a quadratic equation in the constant "4":

$$4^2 - (2x^2 + 1)4 + (x^4 - x) = 0.$$

Using the quadratic formula on 4, we get

$$4 = \frac{2x^2 + 1 \pm \sqrt{(2x^2+1)^2 - 4(x^4 - x)}}{2} = \frac{2x^2 + 1 \pm (2x+1)}{2}.$$

So $4 = x^2 + x + 1$ or $4 = x^2 - x$. Solve and throw away the negative roots, $x = \dfrac{\sqrt{13} - 1}{2}$ or $x = \dfrac{\sqrt{17} + 1}{2}$. Further check shows that $x = \dfrac{\sqrt{13} - 1}{2} < 2$ is also extraneous. So there is only one root $x = \dfrac{\sqrt{17} + 1}{2}$. (It might help for easier understanding to do it this way: Let $y = 4$: $\sqrt{\sqrt{x+y} + y} = x$, and then solve for y in terms of x.)

Problem 8.6 Solve for real x: $\sqrt{\dfrac{x-2}{x+2}} + \sqrt{\dfrac{9x+18}{x-2}} = 4$.

Answer

$\dfrac{-5}{2}$.

Solution

Make the substitution $y = \sqrt{\dfrac{x-2}{x+2}}$ to get $y + \dfrac{3}{y} = 4$. Solving for y gives $y = 1, 3$. Hence $\dfrac{x-2}{x+2} = 1$ or 9. 1 is impossible, so $x - 2 = 9(x+2)$ and hence $x = -5/2$.

Problem 8.7 Solve for x: $(x - \sqrt{3})x(x+1) + 3 - \sqrt{3} = 0$.

Answer

$x = \sqrt{3} - 1$ and $x = \pm\sqrt[4]{3}$.

Solution

This is a cubic equation, and it is difficult to solve directly. So we let $y = \sqrt{3}$. Then $(x - y)x(x + 1) + y^2 - y = 0$. Expand the left hand side:

$$x^3 - x^2y + x^2 - xy + y^2 - y = 0.$$

Treat this equation as a quadratic equation in y,

$$y^2 - (x^2 + x + 1)y + (x^3 + x^2) = 0,$$

which is

$$y^2 - (x^2 + x + 1)y + x^2(x + 1) = 0,$$

and this is clearly factored as

$$(y - x - 1)(y - x^2) = 0,$$

so we have $y = x + 1$ and $y = x^2$. Thus the solutions: $x = \sqrt{3} - 1$ and $x = \pm\sqrt[4]{3}$.

Problem 8.8 Let a be a real number, and the equation $x^2 + a^2x + a = 0$ has real roots for x. Find the maximum possible root x.

Answer

$x_{\max} = \sqrt[3]{2}/2$.

Solution

$x_{\max} = \sqrt[3]{2}/2$. See the equation as a quadratic equation in a, $xa^2 + a + x^2 = 0$; the discriminant is $1^2 - 4x^3 \geq 0$, so $4x^3 \leq 1$, thus $x \leq \dfrac{1}{\sqrt[3]{4}} = \dfrac{\sqrt[3]{2}}{2}$.

Problem 8.9 Solve $\sqrt{2x+2} - \sqrt{x+3} = \sqrt{x+1} - \sqrt{2x+4}$.

Answer

$x = -1$.

Solution

Square both sides. After simplifying you get $\sqrt{(2x+2)(x+3)} = \sqrt{(x+1)(2x+4)}$ so $(2x+2)(x+3) = (x+1)(2x+4)$. Expanding and combining like terms we get $2x = -2$ so $x = -1$. Double checking, it is a solution.

Problem 8.10 For what range of k does $\sqrt{2x^2+4} = x+k$ have real solutions?

Answer

$k \geq \sqrt{2}$.

Solution

Squaring both sides we have $2x^2 + 4 = x^2 + 2kx + k^2$ so $x^2 - 2kx + 4 - k^2 = 0$. The discriminant is

$$\Delta = 4k^2 - 4(4 - k^2) = 8k^2 - 16.$$

Hence the equation has real roots when $8k^2 \geq 16$ so $k^2 \geq 2$, or $k \geq \sqrt{2}$ or $k \leq -\sqrt{2}$. However, if $k \leq -\sqrt{2}$ we have extraneous roots, so $k \geq \sqrt{2}$. (It may belp to think of this part graphically.)

9   Solutions to Chapter 9 Examples

Problem 9.1 For each of the following statements, say whether they are true or false. If false, give a counterexample and correct the statement.

(a) Every linear function has an inverse.

Answer

False.

Solution

The linear function $y = k$ (for a constant k) does not have an inverse. However, every other linear function (with slope $\neq 0$).

(b) Every (non-constant) polynomial function $f(x)$ can be divided into pieces so that each piece has an inverse.

Answer

True

Solution

Divide it into pieces where it is either increasing or decreasing. For example, a quadratic can be divided into two parts on either side of the vertex.

(c) If $f(g(x)) = x$, then $g(f(x)) = x$.

Answer

False.

Solution

Let $f(x) = x^2$ and $g(x) = \sqrt{x}$. Then $f(g(x)) = x$, but $g(f(x)) = |x|$.

Problem 9.2 Find the inverse of the following functions.

(a) $f(x) = 4x + 3$.

Answer

$$\frac{x-3}{4}.$$

Solution

Solve $x = 4y + 3$ for y to get the inverse.

(b) $f(x) = \dfrac{1}{x+1}.$

Answer

$$\frac{x-1}{x}.$$

Solution

Solve $x = \dfrac{1}{y+1}$ for y to get the inverse.

(c) $f(x) = x + \dfrac{1}{x}$ for $x \geq 1$. Note: $f(1) = 2$ is the minimum of $f(x)$.

Answer

$$\frac{x + \sqrt{x^2 - 4}}{2} \text{ for } x \geq 2.$$

Solution

Solve $x = y + \dfrac{1}{y}$ for y to get the inverse. Note you need to solve a quadratic in y to get $y = (x \pm \sqrt{x^2 - 4})/2$ for $x \geq 2$. Since $x \geq 1$ we only need the $+$ version, so $f^{-1}(x) = \dfrac{x + \sqrt{x^2 - 4}}{2}$ for $x \geq 2$.

Problem 9.3 Assume $x > 0$. Let $f(x) = x^2$, $g(x) = 2x + 3$, and $h(x) = \dfrac{1}{x}$. Calculate:

(a) $f^{-1}(g(h(2)))$.

Answer

2.

Solution

$h(2) = 1/2$, and $g(1/2) = 4$. Since $f(x) = x^2$ and $x > 0$, we have that $f^{-1}(x) = \sqrt{x}$. Thus $f^{-1}(4) = 2$.

(b) $h(g^{-1}(f^{-1}(36)))$

Answer

$\dfrac{2}{3}$.

Solution

Since $f(x) = x^2$ and $x > 0$, we get $f^{-1}(x) = \sqrt{x}$. So $f^{-1}(36) = 6$.

Since $g(x) = 2x + 3$, $g^{-1}(x) = \dfrac{x-3}{2}$, thus $g^{-1}(6) = \dfrac{3}{2}$. Thus

$$h\left(\frac{3}{2}\right) = \frac{2}{3}.$$

(c) $h(h(f(h(h(g(g^{-1}(g(x))))))))$.

Answer

$4x^2 + 12x + 9$.

Solution

Note $h^{-1}(x) = h(x)$, so the above becomes $f(g(x))$ after canceling inverses. Thus
$$h(h(f(h(h(g(g^{-1}(g(x)))))))) = f(g(x)) = (2x+3)^2 = 4x^2 + 12x + 9.$$

Problem 9.4 Prove $\log_b(x) + \log_b(y) = \log_b(x \cdot y)$ using the rules for exponents and the definition of logs as the inverse of the exponential function.

Solution

The above is true if and only if $b^{\log_b(x) + \log_b(y)} = x \cdot y$. Using rules of exponents and the fact that $\log_b(z)$ is the inverse of b^z
$$b^{\log_b(x) + \log_b(y)} = b^{\log_b(x)} \cdot b^{\log_b(y)} = x \cdot y$$

as needed.

Problem 9.5 Calculate the following

(a) $\log_5(125)$.

Answer

3.

Solution

Note $5^3 = 125$.

(b) $\log_2(8) + \log_{\sqrt{2}}(4)$

Answer

7.

Solution

Note $2^3 = 8$ and $\sqrt{2}^4 = 4$ so our answer is $3 + 4 = 7$.

(c) $\log(30) - \log(3)$.

Answer

1

Solution

Using our rules for logarithms, $\log(30) - \log(3) = \log(30/3) = \log(10) = 1$. (Recall $\log = \log_{10}$.)

(d) $\dfrac{\log_2(343)}{\log_2(49)}$.

Answer

$\dfrac{3}{2}$.

Solution

Using the change of base formula, $\dfrac{\log_2(343)}{\log_2(49)} = \log_{49}(343) = 3/2$ as $49 = 7^2$ and $343 = 7^3$ so $49^{3/2} = 343$.

Problem 9.6 What is the domain and range of $\log_3(\log_{1/3}(x+2))$? What are its zeros?

Answer

Domain: $-2 < x < -1$; Range: All Reals; Zeros: $x = -5/3$.

Solution

$\log_{1/3}(x+2)$ is defined as long as $x > -2$. For $\log_3(y)$ to be defined, $y > 0$ so $\log_{1/3}(x+2) > 0$. Hence $\log_{1/3}(z) > 0$ as long as $0 < z < 1$, so $x+2 < 1$ and $x < -1$. Combining the domain is $-2 < x < -1$. The range is all real numbers and $\log_3(\log_{1/3}(x+2)) = 0$ if $\log_{1/3}(x+2) = 1$ which happens when $x+2 = 1/3$ or $x = -5/3$.

Problem 9.7 Solve the equation $\log_2(x) + \log_4(x) + \log_8(x) = \dfrac{22}{3}$.

Answer

$x = 16$.

Solution

We have (using change of base)

$$\begin{aligned}
\log_2(x) + \log_4(x) + \log_8(x) &= \log_2(x) + \frac{\log_2(x)}{\log_2(4)} + \frac{\log_2(x)}{\log_2(8)} \\
&= \log_2(x) + \frac{\log_2(x)}{2} + \frac{\log_2(x)}{3} \\
&= \frac{11\log_2(x)}{6}.
\end{aligned}$$

Hence we need $\log_2(x) = 4$ so $x = 2^4 = 16$.

Problem 9.8 For how many integer values is $\log(x-20) + \log(30-x) < 1$?

Answer

2

Solution

Note first that the domain of $\log(x-20)+\log(30-x)=\log[(x-20)(30-x)$ is $20 < x < 30$. Trying $x = 21$ we have $\log(9) < 1$ but if $x = 22$ we have $\log(16) > 1$. By symmetry, we have if $x = 28$ does not work, but $x = 29$ does work. Hence there are two integer solutions.

Problem 9.9 Solve $2\log_2(x)\log_4(x)+2\log_2(x)-3=0$.

Answer

$x = 2, x = 1/8$.

Solution

Note $\log_4(x)=\log_2(x)/2$ so the equation is $\log_2^2(x)+2\log_2(x)-3=0$. Let $z = \log_2(x)$ so we have $z^2+2z-3=(z+3)(z-1)=0$ or $z = -3, z = 1$. If $z = 1$ we have $x = 2$ and if $z = -3$ we have $x = 1/8$. Both are solutions.

Problem 9.10 (AIME 1983) Let x, y, and z all exceed 1, and let w be a positive number such that $\log_x(w) = 24$, $\log_y(w) = 40$, and $\log_{xyz}(w) = 12$. Find $\log_z(w)$.

Answer

60.

Solution

First rewrite as $\log_w(x) = 1/24$, $\log_w(y) = 1/40$, $\log_w(xyz) = 1/12$. Since $\log_w(xyz) = \log_w(x) + \log_w(y) + \log_w(z)$ we have $\log_w(z) = 1/12 - 1/24 - 1/40 = 1/60$. Hence $\log_z(w) = 60$.

www.ingramcontent.com/pod-product-compliance
Lightning Source LLC
Chambersburg PA
CBHW081507200326
41518CB00015B/2408